Fred J. Hickernell, Peter Kritzer (Eds.)

Multivariate Algorithms and Information-Based Complexity

Radon Series on Computational and Applied Mathematics

Volume 27

Multivariate Algorithms and Information-Based Complexity

Edited by
Fred J. Hickernell and Peter Kritzer

DE GRUYTER

Editors
Prof. Dr. Fred J. Hickernell
Illinois Institute of Technology
Department of Applied Mathematics
Center for Interdisciplinary Scientific
3105 Dearborn St.
Chicago 60616
USA
peter.kritzer@oeaw.ac.at

PD Dr. Peter Kritzer
Austrian Academy of Sciences
Altenbergerstraße 69
4040 Linz
Austria
peter.kritzer@oeaw.ac.at

ISBN 978-3-11-063311-5
e-ISBN (PDF) 978-3-11-063546-1
e-ISBN (EPUB) 978-3-11-063315-3
ISSN 1865-3707

Library of Congress Control Number: 2020934321

Bibliographic information published by the Deutsche Nationalbibliothek
The Deutsche Nationalbibliothek lists this publication in the Deutsche Nationalbibliografie; detailed bibliographic data are available on the Internet at http://dnb.dnb.de.

Typesetting: VTeX UAB, Lithuania
Printing and binding: CPI books GmbH, Leck

www.degruyter.com

Preface: Multivariate algorithms and information-based complexity

The authors of this book include several of the invited speakers in the workshop *Multivariate Algorithms and Information-Based Complexity*, which was part of the RICAM Special Semester on *Multivariate Algorithms and their Foundations in Number Theory* in the fall of 2018. The special semester consisted of four larger and two smaller workshops on various topics ranging from Pseudo-Randomness and Discrepancy Theory to Information-Based Complexity and Uncertainty Quantification. This book arises from the second workshop, which took place at the Johann Radon Institute for Computational and Applied Mathematics (RICAM) of the Austrian Academy of Sciences in Linz, Austria, on November 5–9, 2018.

Multivariate continuous problems occur in a multitude of practical applications, such as physics, finance, computer graphics, and chemistry. The number of variables involved, d, can be in the hundreds or thousands. The information complexity of a given problem is the minimal number of information operations required by the best algorithm to solve the problem for a prescribed set of inputs within a certain error threshold, ε. Typical examples of information operations are function values and linear functionals. The field of information-based complexity (IBC), founded by Traub and Wozniakowski in the 1980s, analyzes the information complexity for multivariate problems and determines how it depends on d and ε. A crucial question is under which circumstances one can avoid a curse of dimensionality, namely, exponential growth of the information complexity with d. This book addresses the analysis of multivariate (continuous) problems, especially from the IBC viewpoint.

The problems discussed by the authors reflect the breadth of current inquiry under the umbrella of multivariate algorithms and IBC. The chapter entitled"The control variate integration algorithm for multivariate functions defined at scattered data points" studies a method of approximating the integral of a multivariate function, in which one uses the exact integral of a control variate based on a least-squares multivariate quasiinterpolant. Numerical examples demonstrate that such an algorithm can overcome the curse of dimensionality for multivariate least-squares fits. The second chapter, titled "An adaptive random bit multilevel algorithm for SDEs", considers the approximations of expectations for functionals applied to the solutions of stochastic differential equations by employing Monte Carlo methods based on random bits instead of random numbers. An adaptive random bit multilevel algorithm is provided and compared numerically to other methods. The chapter "RBF-based penalized least-squares approximation of noisy scattered data on the sphere" deals with the approximation of noisy scattered data on the 2-dimensional unit sphere. In particular, global and local penalized least-squares approximation based on radial basis functions (RBFs) are explored. The authors of the fourth chapter in this book, "On the power of random information", consider a problem from the core of IBC theory,

https://doi.org/10.1515/9783110635461-201

namely the quality of random information in approximation and integration problems. Random information is compared to optimal information for such problems. It is shown that in some cases random information can be almost optimal, whereas in other cases it may yield much worse results than optimal information. The chapter "Optimality criteria for probabilistic numerical methods" lies in the field of Bayesian decision theory. To be more precise, the authors study an optimality criterion from Bayesian experimental design and consider its implied optimal information in the numerical context. Furthermore, this chapter compares this information to information commonly used in average-case-optimal numerical methods. The sixth chapter of the book, "ε-Superposition and truncation dimensions, and multivariate decomposition method for ∞-variate linear problems" deals with linear problems on weighted normed function spaces, the elements of which depend on infinitely many variables. The author focuses on the ε-truncation and ε-superposition dimension, as well as multivariate decomposition methods, which are means to reduce the number of variables in the problem to a relatively small finite number. The anchored and other types of decompositions are used. The chapter "Adaptive approximation for multivariate linear problems with inputs lying in a cone" completes the book and considers adaptive approximation algorithms for functions lying in particular subsets of certain function spaces. In contrast to much of the IBC literature, it is not assumed that the functions to be approximated lie in a ball of a certain radius, but instead in a cone. It is shown that for such problems it is an advantage to use adaptive rather than nonadaptive algorithms for function approximation.

All chapters were reviewed by renowoned experts. We wish to thank these anonymous referees for their precious help.

We would like to express our gratitude towards all speakers of the workshop for giving excellent talks on their respective fields of expertise and for contributing to the success of the workshop:
- Paul Constantine (University of Colorado),
- Martin Ehler (University of Vienna),
- Mario Hefter (TU Kaiserslautern),
- Stefan Heinrich (TU Kaiserslautern),
- Kerstin Hesse (Paderborn University),
- Fred J. Hickernell (Illinois Institute of Technology),
- James (Mac) Hyman (Tulane University),
- Thomas Müller-Gronbach (University of Passau),
- Erich Novak (University of Jena),
- Chris Oates (Newcastle University and Alan Turing Institute),
- Houman Owhadi (California Institute of Technology),
- Sergei Pereverzyev (RICAM), Leszek Plaskota (University of Warsaw),
- Paweł Przybyłowicz (AGH University of Science and Technology Kraków),
- Klaus Ritter (TU Kaiserslautern),
- Ingo Steinwart (University of Stuttgart),

- Michaela Szölgyenyi (University of Klagenfurt),
- Greg W. Wasilkowski (University of Kentucky),
- Henryk Woźniakowski (Columbia University and University of Warsaw),
- Henry Wynn (London School of Economics),
- Larisa Yaroslavtseva (University of Passau).

We also would like to thank all those who participated in the workshop. Annette Weihs and Wolfgang Forsthuber provided valuable administrative and technical support. Further information on the RICAM Special Semester "Multivariate Algorithms and their Foundations in Number Theory" can be found at

https://www.ricam.oeaw.ac.at/specsem/specsem2018/

The financial support of the Johann Radon Institute for Computational and Applied Mathematics (RICAM) of the Austrian Academy of Sciences is gratefully acknowledged.

Fred J. Hickernell and Peter Kritzer,
Chicago and Linz, March 2020.

Contents

Jeremy Dewar, Mu Tian, and James M. Hyman

1 The control variate integration algorithm for multivariate functions defined at scattered data points

Abstract: We describe numerical studies to demonstrate the accuracy and efficiency of approximating the integral of a function over a prescribed multidimensional region using the exact integral of a control variate based on a least-squares multivariate polynomial quasiinterpolant. This control variate integration approach is shown to be effective in approximating the integral of smooth functions, where only a few function values are available. We compare these methods with traditional Monte Carlo quadrature on independent and identically distributed random and low-discrepancy (quasi-Monte Carlo) points. We verify that the control variate integration methods converge at the same rate as Monte Carlo methods, but have a much smaller error constant. We provide numerical examples to demonstrate that sparse quasiinterpolation methods are a practical approach to reduce the curse of dimensionality for high-dimensional multivariate least-squares fits.

Keywords: Quadrature, cubature, multivariate, integration, low discrepancy sampling, quasi-Monte Carlo

MSC 2010: 62H12, 65D30, 65D05

1.1 Introduction

Estimating the integrals of high-dimensional multivariate functions arises in estimating the marginal likelihood in Bayesian analysis of problems ranging from quantifying the uncertainty in pricing financial options [3, 15, 22] to the solution of stochastic partial differential equations with random coefficients, such as the flow of fluids through porous media [4, 18]. There are accurate and robust numerical methods to integrate a function over a low-dimensional domain. Most of these methods are either built on a lattice grid or a particular set of deterministic points, such as Gaussian or Smolyak

Acknowledgement: The authors thank Lin Li for his feedback on these simulations and many helpful discussions about the CVI and future extensions of the method. We are grateful to Chris Oates for his many helpful suggestions to improve the paper. This research was partially supported by the funding from Tulane University Evelyn and John G. Phillips Distinguished Chair in the Mathematical Sciences.

Jeremy Dewar, James M. Hyman, Department of Mathematics, Tulane University, New Orleans, LA 70118, USA, e-mails: jdewar@tulane.edu, mhyman@tulane.edu
Mu Tian, Facebook, 1101 Dexter Ave N, Seattle, WA 98109, USA, e-mail: mtian@fb.com

https://doi.org/10.1515/9783110635461-001

quadrature points, and are based on the exact integration of an interpolating function through the sample points.

As the dimension of the problem increases, traditional numerical integration algorithms [6] suffer from the curse of dimensionality, and require the number of samples to increase exponentially. Integration formulas built on sparse grids [2, 21] partially overcame the curse of dimensionality. However, even these approaches suffer from the curse of dimensionality above eight to ten dimensions. Monte Carlo (MC) integration is based on independent identically distributed (IID) pseudo-random sampling and the mean of the samples converges, with probability 1, to the desired integral by the strong law of large numbers. The MC has a probabilistic convergence rate of $O_p(1/\sqrt{N})$, independent of the dimension, as the number of samples, N, increases.

The quasi-Monte Carlo (QMC) method is based on nonrandom low-discrepancy sampling (LDS) and converges faster, $O((\log N)^{(d-1)}/N)$, where d is the dimension [3, 9]. Recent improvements in the generating functions for low-discrepancy sampling has improved the asymptotic convergence rates of QMC methods [2, 8, 19]. Unfortunately, the errors of the best QMC methods can still be unacceptably large when there are few samples.

For higher-dimensional situations when there are few samples, $f(\mathbf{x}_i)$, available, we fit a least-squares quasiinterpolant, $\hat{f}(\mathbf{x})$, through the data that minimizes the variance of the residuals, $r_i = \hat{f}(\mathbf{x}_i) - f(\mathbf{x}_i)$. This least squares fit is a *quasiinterpolant,* since it is not exact at the data points and is a control variate estimating the underlying function. We then approximate the integral of $f(x)$ by the exact integral of $\hat{f}(\mathbf{x})$ and an estimate for the integral of the residual. We demonstrate that this *control-variate integraton*, (CVI), algorithm is an effective approach [12] for smooth functions, where there are few samples, and the effective dimension is relatively small. This is often the case when additional function evaluations (samples), $f(\mathbf{x})$, are not available or expensive, such as with large-scale simulations.

The MC and QMC methods approximate the integral of the underlying function by the mean of the samples. This approximation is equivalent to fitting the data with a least squares constant and integrating this constant. That is, traditional Monte Carlo integration can be viewed as a CVI using a least-squares constant function. We observed that the higher-degree quasiinterpolants could reduce this error by a factor of 10, 100, or more.

Let $\mathbf{x} \in R^d$ and the function $f(\mathbf{x}) : R^d \rightarrow R^1$ be integrable and bounded. The integral of $f(\mathbf{x})$ on the d-dimensional unit hypercube $\Omega = [0,1]^d$ is

$$\int_{[0,1]^d} f(\mathbf{x})d\mathbf{x} = \int_0^1 \cdots \int_0^1 f(x_1, x_2, \ldots, x_d)dx_1 \cdots dx_d\,. \tag{1.1}$$

All integrations in this paper will be over the domain $[0,1]^d$, and the domain will be dropped from the integral notation from here on. When the integral is over a different region, we will assume that the problem can be mapped to the unit cube through a change of variables. Note that creating a good mapping can be a nontrivial problem [16]. Even in bounded regions with large aspect ratios, the mapping can convert a function that is equally smooth in all directions to one that can vary much more in some directions than others.

A traditional approach for approximating the integral is by weighted sum [7],

$$I_f \stackrel{\text{def}}{=} \int f(\mathbf{x})d\mathbf{x} \approx \sum_{i=1}^{N} w_i f(\mathbf{x}^i) \stackrel{\text{def}}{=} \hat{I}_f\,, \tag{1.2}$$

where w_i is the integration weight associated with the sample point $f(\mathbf{x}^i)$, and $\mathbf{x}^i \in [0,1]^d$ for $i = 1,\ldots,N$ samples. If the samples are approximately uniformly distributed in $[0,1]^d$, then the Monte Carlo methods use the sample mean, $w_i = 1/N$, to approximate the integral. Most classical integration methods define the weights so that the quadrature is exact for a class of functions interpolating the data.

The asymptotic integration error estimates for the MC method is

$$|\hat{I}_f - I_f| = O_p\left(\frac{1}{\sqrt{N}}\sigma(f)\right), \tag{1.3}$$

where $\sigma(f) = [\int (f(\mathbf{x}) - I_f)^2 dx]^{1/2}$. The corresponding asymptotic integration error estimate for QMC method is

$$|\hat{I}_f - I_f| = O\left(\frac{(\log N)^{(d-1)}}{N} V[f]\right). \tag{1.4}$$

Here, $V[f]$ is the variation in f, $V[f] = \int |\frac{\partial^d f}{\partial_{x_1}\cdots\partial_{x_d}}| dx_1\cdots dx_d + \sum_{i=1}^{N} V[f(\mathbf{x}^i)]$ [20]. Note that the QMC error is not a tight bound, and in practice, the error often converges faster. Also, although the QMC method converges faster, the error can be biased, unlike the MC method [3].

In the CVI, we first fit the sample with a least-squares quasiinterpolant, $\hat{f}(\mathbf{x})$. We then use this function as a control variate and estimate the integral by the exact integral of $\hat{f}$, plus an approximation of the residual $r(x) = f(x) - \hat{f}(x)$,

$$\int f(\mathbf{x})dx = \int \hat{f}(\mathbf{x})dx + \int f(\mathbf{x}) - \hat{f}(\mathbf{x})dx = \underbrace{\int}_{\text{exact}} \hat{f}(\mathbf{x})dx + \int r(\mathbf{x})dx\,. \tag{1.5}$$

The error is now determined by the integral of the residual and asymptotically decreases at the same rate as the original Monte Carlo approximation. This is to be expected since the error estimates are the product of the sample discrepancy based on the distribution and number of samples, which are unchanged, and the variation of the function being integrated, which is independent of the number of samples.

Since the integral of the control variate, $\int \hat{f}$, is known exactly, the only error is the integral of the residual $\int r(x)dx$. That is, the CVI integration error estimates are the same as the MC and QMC error estimates when applied to integrating the residual $r(\mathbf{x})$, as they were for integrating $f(\mathbf{x})$. The residual is in the null space of the basis functions. Therefore, the expected value is zero, $E(r) = 0$, if the quasiinterpolant is exact for constant functions. If we assume that the integral of the residual is zero, then the Monte Carlo methods can be viewed as a CVI, where the quasiinterpolant is the constant function.

The CVI is also called a regression estimator, and is typically used with low- degree approximating functions [5, 12, 14, 17]. Our numerical study focuses on using higher-degree polynomial control variates and analyzing the magnitude of the error (i. e., the constant in front of that convergence rate) for both a fixed number of samples N, and as the number of samples increases. We also analyze the complexity of the algorithm as the dimension d increases.

1.2 Quasiinterpolation algorithm

We represent our control variates as a sum of orthogonal basis functions, $f(\mathbf{x}) = \sum_k \sum_j \beta_{j,k}\psi_{j,k}(\mathbf{x})$, with total degree K. The multivariate basis functions, $\psi_{j,k}(\mathbf{x}) : R^d \to R^1$ of degree $k \le K$, are decomposed into the product univariate basis functions, $\{\phi_{k_i^j}(x_i)\}_{j=1}^{t_k^d} : R^1 \to R^1$,

$$\psi_{j,k}(\mathbf{x}) = \prod_{i=1}^{d} \phi_{k_i^j}(x_i)\,. \tag{1.6}$$

Here, $\mathbf{x} = [x_1, x_2, \ldots, x_d]^T \in R^d$ and k_i^j is the highest degree of x_i in the univariate basis function $\phi_{k_i^j}(x_i)$, where j ranges from 1 to $t_K^d = \binom{K+d-1}{d-1}$ and is constrained by $\sum_{i=1}^d k_i^j \le K$. In this paper, we use the Legendre polynomial basis functions shifted to the interval $[0, 1]$: $\phi_0(x) = 1$, $\phi_1(x) = 2x - 1$, $\phi_2(x) = 6x^2 - 6x + 1, \ldots$.

A polynomial of total degree K has $M = \sum_{l=1}^K t_l^d = \binom{K+d}{d}$ basis functions. We order and relabel β_l and $\psi_l(\mathbf{x})$ lexicographically as a function of j and k, and express our control variate quasiinterpolant as

$$\hat{f}(\mathbf{x}) = \sum_{l=1}^{M} \hat{\beta}_l \psi_l(\mathbf{x}). \tag{1.7}$$

High-degree polynomials in high-dimensions are quickly constrained by the curse of dimensionality (Table 1.1). For the solution to be unique and the coefficients identifiable, then, there must be more independent samples than coefficients in the quasiinterpolating function. For even moderate degree polynomials in moderate dimensional problems, the number of terms rapidly increases.

Table 1.1: Table of the number of terms in multivariate polynomial with total degree K. The mixed variable terms in multivariate polynomials create an explosion in the number of terms as a function of the degree and dimension. For example, a 5th degree polynomial in 20 dimensions has 53,130 terms. We use sparse quasiinterpolation algorithms to reduce the number of terms in the quasiinterpolant.

K / d =	1	2	3	4	5	10	20
0	1	1	1	1	1	1	1
1	2	3	4	5	6	11	21
2	3	6	10	15	21	66	231
3	4	10	20	35	56	286	1,771
4	5	15	35	70	126	1,001	10,626
5	6	21	56	126	252	3,003	53,130
6	7	28	84	210	462	8,008	230,230
10	11	66	286	1,001	3,003	184,756	30,045,015

The coefficients $\hat{\beta}_l$ are defined by solving the linear ordinary least-squares problem

$$\hat{\boldsymbol{\beta}} = \underset{\beta_1,\beta_2,\ldots\beta_M}{\operatorname{argmin}} \sum_i (\hat{f}(\mathbf{x}^i) - f(\mathbf{x}^i))^2 = \underset{\beta_1,\beta_2,\ldots\beta_M}{\operatorname{argmin}} \|A\boldsymbol{\beta} - \mathbf{f}\|_2^2 . \tag{1.8}$$

Here, $\mathbf{x}^i$ is the ith sample value, and the $(i,j)^{\text{th}}$ element of the design matrix A is defined by $a_{i,j} = \psi_j(\mathbf{x}^i)$.

One of the challenges in least squares approximations is the variance-bias tradeoff. If the expansion has too many high-degree terms, the variance increases and the function will be over-fit. If there are too few terms, then the function will be underfit, and the residual errors will be biased and larger than necessary. We use an l_1 penalty function (LASSO) to regularize the least-squares fit and pick out the most needed terms using sparse subset selection. That is, instead of solving the standard least-squares problem (1.8), we add an l_1 penalty term and solve

$$\hat{\boldsymbol{\beta}} = \underset{\beta_1,\beta_2,\ldots\beta_M}{\operatorname{argmin}} \left\{ \frac{1}{2} \|A\boldsymbol{\beta} - \mathbf{f}\|_2^2 + \lambda \|\boldsymbol{\beta}\|_1 \right\} . \tag{1.9}$$

We solve the system using cyclic coordinate descent algorithm [10, 13], and the optimal λ is determined by a cross-validation test. If the function $f(\mathbf{x})$ varies in some directions more than others, then this sparse subset selection approach extracts the appropriate basis functions based on the effective dimension of the active subspaces.

We used all the samples when fitting the data and chose λ by the simple method of increasing λ until just below, where there was a significant increase in the norm of the residuals, $\|\hat{y} - y\|$. We will be comparing this approach with more sophisticated cross-validation data-splitting methods in the future.

An advantage of using sparse subset selection is that the quadrature only has to be evaluated on the remaining sparse set basis functions with nonzero coefficients. We also considered using a simpler l_2 regularization for the least squares fit. This results in solving an easier (linear) problem for the control variate coefficients, but requires evaluating the integral of all of the basis functions. It is not clear which of these approaches will be the most effective, and we are currently comparing them on both smooth and nonsmooth test problems.

1.3 Numerical examples

We compare the methods on five- and six-dimensional smooth functions by generating LDS and IID samples form uniform distributions of size $500 \leq N \leq 4000$, and quasiinterpolating polynomials with total degrees $K = 0, 1, \ldots 5$. We then compare the methods a nonsmooth wedge function in six dimensions. We analyze the distributions of errors from an ensemble of 300 runs, each with distinct samples. The IID samples are generated using the MATLAB pseudorandom generator, *rand*, and the LDS are generated using the Niederreiter or Sobol sequences.

For the test problems, we calculate the exact error, the convergence rate, the constant factor in the error estimate, and an approximate integration error using bootstrapping. We compute the error distributions over a large ensemble of simulations to avoid misleading results, where any particular sample set might have unusually large or small errors.

For each case in the ensemble of 300 runs, the testing procedure is to

1. generate the N sample locations $\{\mathbf{x}^i\}_1^N$,
2. evaluate $f(\mathbf{x}^i)$,
3. fit a (possibly regularized) least-squares polynomial $\hat{f}(\mathbf{x})$ of degree K through the sample,
4. define the approximate integral as $\int \hat{f}(\mathbf{x})d\mathbf{x}$,
5. estimate the error using bootstrapping (resampling with replacement) [13], and finally
6. evaluate the true integration error for the CVI, $e = |\int(\hat{f}(\mathbf{x}) - f(\mathbf{x}))dx|$.

After each run, the true integration error is compared with the bootstrapping approximation of the standard error.

We compared the convergence rate of the CVI method on smooth and discontinuous functions defined on IID and LDSs, for N ranging from a few hundred to a few thousand, and polynomial quasiinterpolants of degree less than 5.

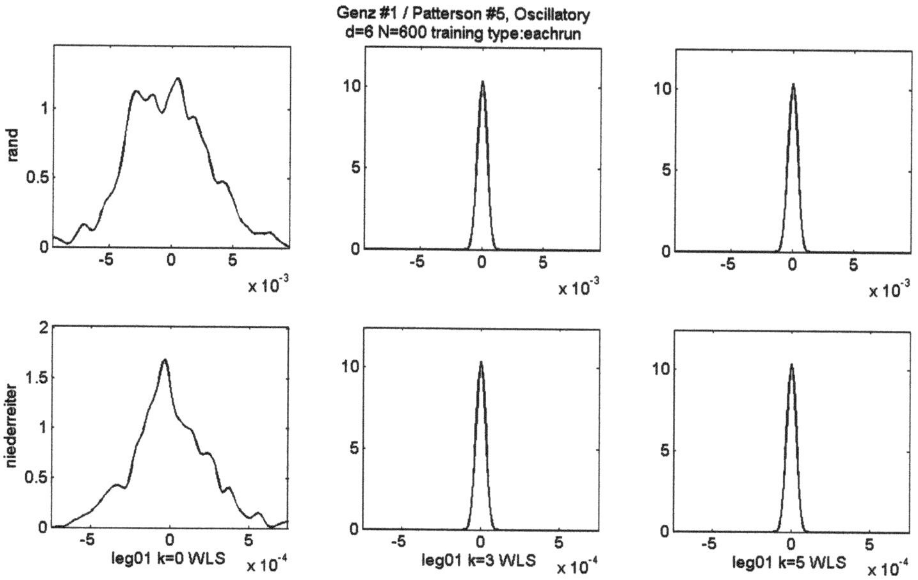

Figure 1.1: The error distributions for integrating $f_g(\mathbf{x})$ in six dimensions, $N = 600$ samples, for quasiinterpolants with degrees $K = 0, 3, 5$. The x-axis for the IID samples (top row) is ten times larger than the x-axis for the LDS samples (bottom row).

1.3.1 Genz oscillatory example 1

We use the Genz oscillatory cosine test function [11],

$$f_g(\mathbf{x}) = \cos\left(0.6\pi + \frac{1}{d}\sum_{i=1}^{d} x_i\right), \tag{1.10}$$

to illustrate the convergence of CVI for smooth functions.

Figure 1.1 shows the error distributions after 300 simulations integrating $f_g(\mathbf{x})$ in six dimensions using 600 samples with MC/QMC ($K = 0$) and the cubic and quintic quasiinterpolant ($K = 3, 5$). In this example, we kept all of the terms in the Kth total degree polynomials (Table 1.1). Notice the errors are approximately unbiased and the x-axis (error) scales are ten times greater for the IID samples than they are for the LDS samples. The standard deviation of $K = 3$ errors are also about ten times smaller than the $K = 0$ errors.

The y-axis in Figure 1.2 is the log of the mean error, and the x-axis is the log of the number of samples $500 \leq N \leq 4000$. The figures show the steady reduction in the errors as the degree of the quasiinterpolant increase from the MC/QMC ($K = 0$) on the left to the quintic-quasiinterpolant on the rightmost figure. The errors for the IID sample distributions (upper blue curves) are always greater than the errors for the LDS samples (lower green curves). This difference is most significant for low-degree quasiinterpolants and is less pronounced for the quintic quasiinterpolant.

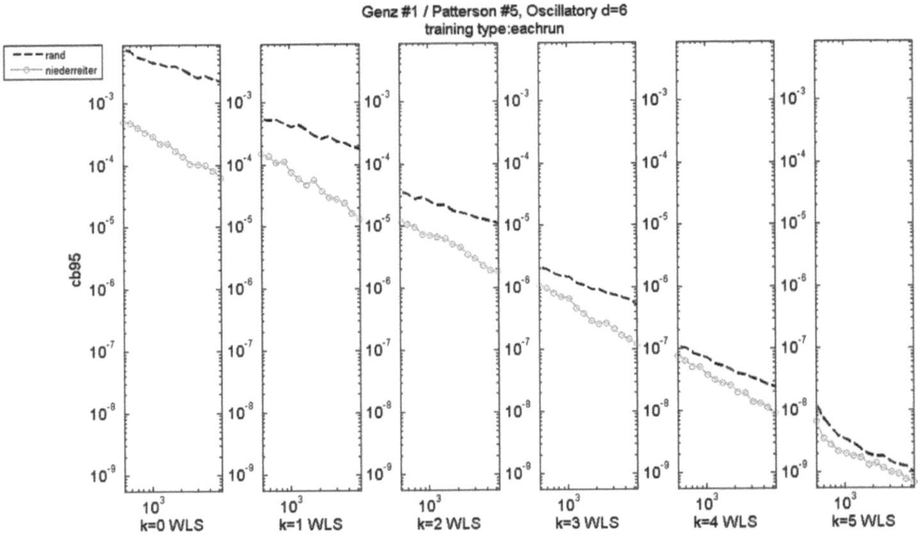

Figure 1.2: The mean integration errors for six dimensional Genz oscillatory function $f_g(\mathbf{x})$ with quasi-interpolating degrees $K = 0, 1, \ldots 5$ for $500 \leq N \leq 4000$ samples. The errors for the IID samples (upper dashed curves) are constantly higher than the error for the LDS samples (lower -o- curves). The left plot for $K = 0$ is equivalent to using the traditional Monte Carlo and quasi-Monte Carlo integration. The convergence rates (Table 1.2) verify that CVI converges approximately as $O_p(1/\sqrt{N})$ for IID and as $O(1/N)$ for LDS samples.

Table 1.2: The convergence rates for integration errors on example (1.10), $f_g(\mathbf{x})$, in six dimensions show that the lower degree CVIs have the same convergence rates as the Monte Carlo methods, but the error constants are greatly reduced. These rates were determined using a sequence of 13 problems with between 500 and 4000 samples. The bootstrap 95 % error confidence bound (ECB) is a good approximation for the accuracy.

IID	$k = 0$ (MC)	$k = 1$	$k = 2$	$k = 3$	$k = 4$	$k = 5$
95 % ECB	2.205e−03	1.785e−04	1.089e−05	5.124e−07	2.399e−08	9.978e−10
α	4.420e−01	5.317e−01	5.015e−01	5.601e−01	6.358e−01	6.833e−01
C	9.77E−02	1.70E−02	7.90E−04	6.59E−05	5.56E−06	3.63E−07

LDS	$k = 0$ (QMC)	$k = 1$	$k = 2$	$k = 3$	$k = 4$	$k = 5$
95 % ECB	6.088e−05	1.326e−05	1.785e−06	1.213e−07	9.211e−09	6.789e−10
α	8.898e−01	1.011e+00	9.340e−01	9.398e−01	8.497e−01	6.611e−01
C	1.33E−01	8.58E−02	5.26E−03	3.80E−04	1.40E−05	2.05E−07

The slope of the mean error curves corresponds to the asymptotic convergence rate. To analyze convergence rates of the mean errors, $\bar{e}_N$, as a function of sample size, N, we assume that the errors have the form $\bar{e}_N \approx CN^{-\alpha}$ and estimate C and α as a function of the degree K. We also calculated the bootstrap 95 % error confidence bound (ECB) for the error.

Note that in Table 1.2, as expected, for low-degree quasiinterpolant, as the number of samples is increased the convergence rates are approximately $O_p(1/\sqrt{N})$ for the IID, and $O(1/N)$ for the LDS samples. The slope of the error curves for the lower-degree CVIs are about the same as for the Monte Carlo methods ($K = 0$). That is, the CVI does not change the asymptotic convergence rate. The improvement is due to reducing the variance of the residuals. This is to be expected, since the errors being reduced in equations (1.3) and (1.4), $\sigma(f)$ and $\mathrm{Var}(f)$, are independent of N. Similar results are illustrated in Figure 5 of [14] for lower- degree control variates.

For the fifth degree quasiinterpolant, the convergence rates on the IID and LDS samples are about the same. The increased accuracy from using CVI comes from reducing the error constant from the variation in the original function $f(\mathbf{x})$ to the residual $r(\mathbf{x})$. That is, the CVI error constants are $C \approx \sigma(r)$ for IID samples and $C \approx V[r]$ for LDSs.

1.3.2 Sum of cosines sparse interpolant

To investigate the impact of using a sparse quasiinterpolant, we integrate the smooth sum of cosines function

$$f_c(\mathbf{x}) = \sum_{i=1}^{d} i\cos(ix_i) \; . \tag{1.11}$$

Figure 1.3 shows the distribution of errors in six dimensions with 600 samples for $K = 0, 3, 5$. The x-axis scaling is ten times smaller for the LDS samples than the IID samples, and the errors are significantly reduced as the degree of the quasiinterpolant is increased. The convergence rates for the errors are similar to the ones shown in Table 1.2. Note that the errors for the cubic and quintic sparse quasiinterpolant are biased when using a LDS. This bias might be the result of correlations in the ensemble of 300 LDSs generated for this example.

The bootstrap error estimates are compared with the true errors and worst-case error estimates for the CVI in Figure 1.4. These error estimates are excellent for all of the LDS CVIs. However, for the IID samples, the bootstrap error estimates are more accurate for the higher-degree quasiinterpolants than they are for the lower-degree ones.

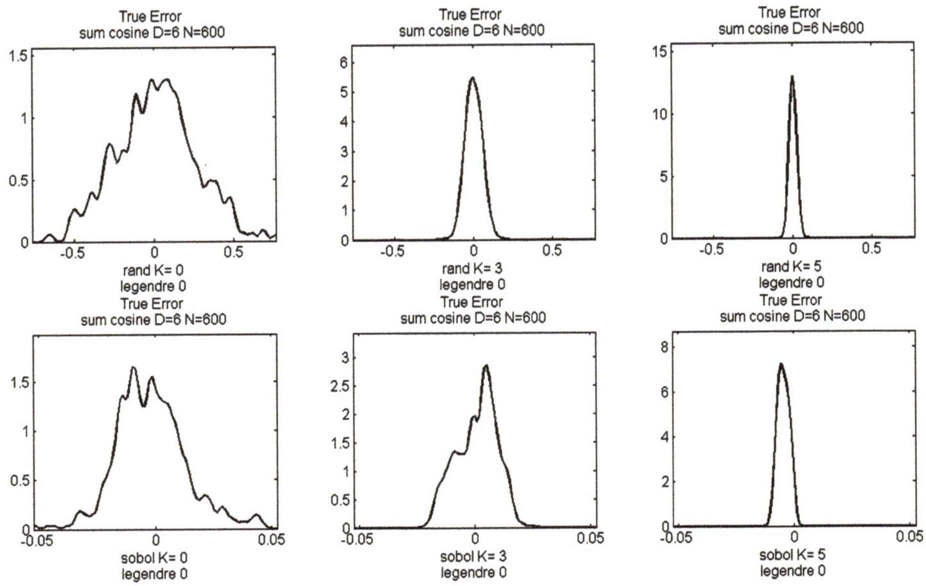

Figure 1.3: The error distributions (6D, 600 samples) for the CVI approximation on IID (top) and LDS (bottom) samples with $K = 0, 3, 5$ degree polynomials. The x-axis bounds are 10 times smaller for the LDS/QMC samples. Note the biased error for the LDS sampling. There are some caveats to using the QMC method; although it converges faster, the error can be biased, unlike the MC method [3]. Quasi-Monte Carlo method uses a non-IID low- discrepancy sample (LDS) and converges faster.

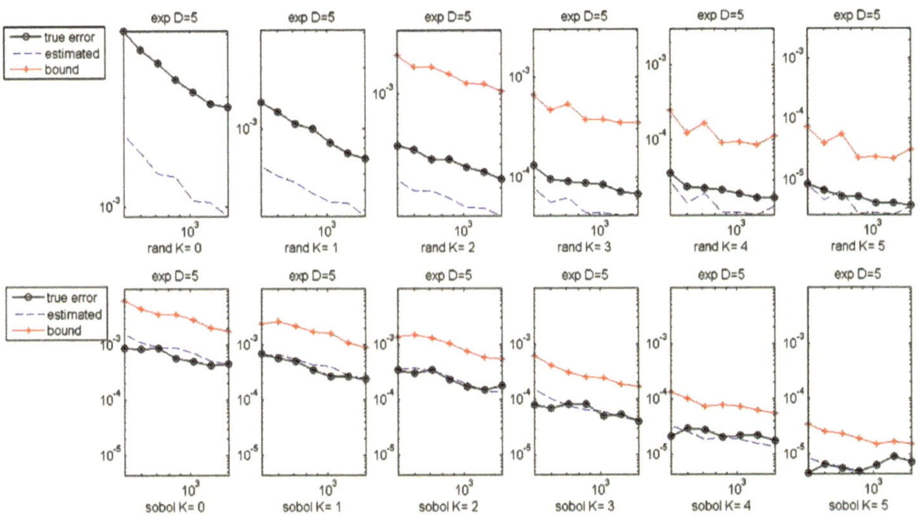

Figure 1.4: The bootstrap error estimates – – – are close to the true errors –∘–∘–, for example, (1.11), $\int f_c(\mathbf{x})dx$. MC (top) and QMC (lower) plots of the worst-case error bounds (red), true error (black), cross-validation error estimates (dashed). All the errors are below the worst-case error estimates, and cross-validation error estimates are better for QMC than MC.

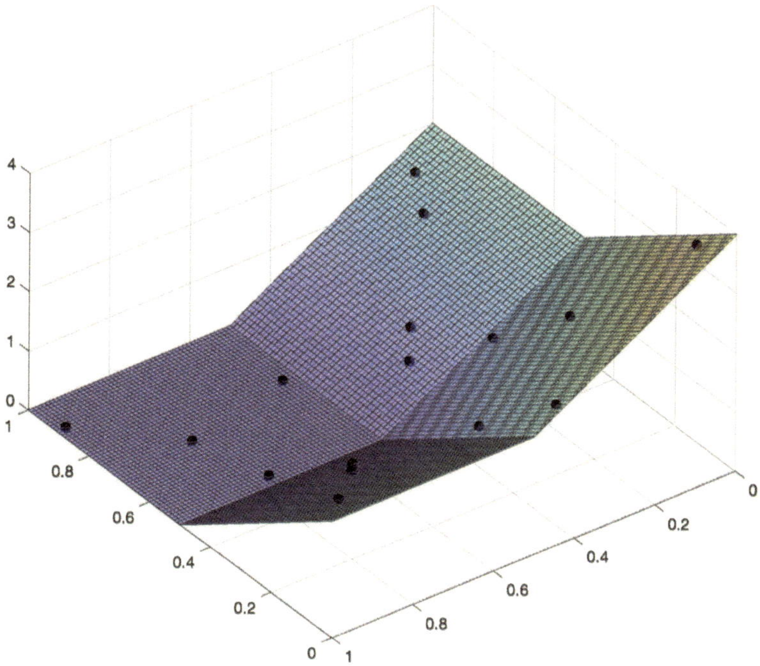

Figure 1.5: The wedge function (1.12) in two dimensions is composed of 2^d planar surfaces.

1.3.3 Wedge function

To investigate how CVI performs on a nondifferentiable C_0 function with corners, we integrated the multidimensional composition of planar surfaces in $[0, 1]^d$,

$$f_w(\mathbf{x}) = \sum_i \max(0, 0.5 - x_i) . \tag{1.12}$$

This wedge function is shown in two dimensions in Figure 1.5.

The mean errors in the six-dimensional example (Figure 1.6) show that the errors on a LDS are consistently smaller than on the IID samples. Although using a linear control variate does slightly improve the errors, the higher degree quasi-interpolants do not significantly reduce the errors. This example is motivating our current research to use local quasiinterpolants, such as distance-weighted least squares and kriging, to better account for local variations in the objective function.

1.4 Conclusions and discussions

Using polynomial quasiinterpolant as control variates is an effective way to reduce errors when integrating smooth high-dimensional functions. We analyzed the effec-

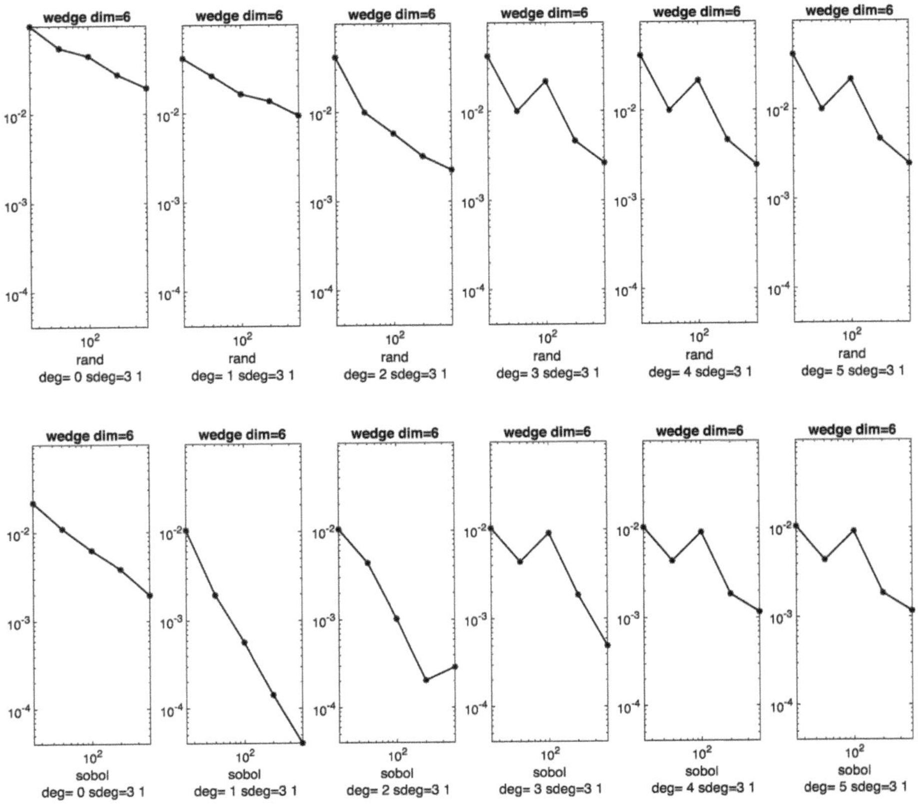

Figure 1.6: High-degree quasiinterpolating does not improve the accuracy of integrating the 6-dimensional wedge function. Errors for MC (top) and QMC (bottom) with quasi-interpolating degree $0 \leq K \leq 5$ for the nonsmooth wedge function (composition of planes).

tiveness of polynomial control variates when there are only a limited number of samples available and the dimension is not too large, $d < 7$. We demonstrated that using sparse polynomial quasiinterpolants for the CVI can help mitigate the explosion in the number of basis functions for high-dimensional problems without compromising the accuracy.

We observed that for smooth functions, the increased accuracy is achieved by reducing the error constant, not the asymptotic error convergence rate. As the number of samples N increased, we observed that the CVI maintained the original Monte Carlo convergence rates of approximately $O_p(1/\sqrt{N})$ and $O(1/N)$ for IID and LDS sample distributions for polynomials of degree less than four ($K < 4$). We also noticed that, for smooth functions, the significant advantage of LDS distributions was much less when using a very high ($K > 3$) quasiinterpolants.

Although the least-squares CVI can be applied to any distribution of samples, we observed that the errors are significantly smaller for LDSs than IID samples. However, the CVI reduced the errors for IID samples more than they did for the LDSs. This result is consistent with previous studies using control variates [1, 14]. We are currently investigating the effectiveness of using CVI on LDS Chebyshev distributions and comparing the results with LDS uniform samples. The hope is that the Chebyshev distributions will better control the quasiinterpolating polynomials.

We also observed that when the underlying function is not smooth, then the errors were somewhat reduced using a linear quasiinterpolant, but using high-degree quasiinterpolants did not significantly reduce, or increase, the accuracy of the CVI. It is relatively easy to construct an example, where a high-degree CVI does worse than pure MC or QMC integration. We will be investigating if the bootstrap error estimates can identify these situations.

We demonstrated that using a sparse control variate based on l_1 regularization is an effective approach for a simple six-dimensional example. The l_1 penalty function both regularized the underdetermined linear system and provided a control mechanism for the bias-variance trade-off of the quasiinterpolant.

Bibliography

[1] H. B. Ameur, P. L'Ecuyer, and C. Lemieux. Variance reduction of Monte Carlo and randomized quasi-Monte Carlo estimators for stochastic volatility models in finance. In *WSC'99. 1999 Winter Simulation Conference Proceedings. 'Simulation-A Bridge to the Future' Cat. No. 99CH37038)*, volume 1, pages 336–343. IEEE, 1999.

[2] H.-J. Bungartz and M. Griebel. Sparse grids. *Acta Numerica*, 13:147–269, 5 2004.

[3] R. E. Caflisch. Monte Carlo and quasi-Monte Carlo methods. *Acta Numerica*, 7:1–49, 1 1998.

[4] J. Charrier, R. Scheichl, and A. Teckentrup. Finite element error analysis of elliptic PDEs with random coefficients and its application to multilevel Monte Carlo methods. Preprint 2/11, University of Bath, March 2011.

[5] W. G. Cochran. Sampling techniques-3. 1977.

[6] R. Cools. Constructing cubature formulae: the science behind the art. *Acta Numerica*, 6:1–54, 1 1997.

[7] P. J. Davis and P. Rabinowitz. *Methods of Numerical Integration: Second Edition*. Dover Publications, November 2007.

[8] F.-M. De Rainville, C. Gagné, O. Teytaud, and D. Laurendeau. Evolutionary optimization of low-discrepancy sequences. *ACM Transactions on Modeling and Computer Simulation*, 22(2):9:1–9:25, March 2012.

[9] J. Dick, F. Y. Kuo, and I. H. Sloan. High-dimensional integration: the quasi-Monte Carlo way. *Acta Numerica*, 22:133–288, 2013.

[10] J. Friedman, T. Hastie, and R. Tibshirani. Regularization paths for generalized linear models via coordinate descent. *Journal of Statistical Software*, 33(1):1, 2010.

[11] A. Genz. Testing multidimensional integration routines. In *Proc. of International Conference on Tools, methods and languages for Scientific and Engineering Computation*, pages 81–94. Elsevier North-Holland, Inc., 1984.

[12] P. W. Glynn and R. Szechtman. Some new perspectives on the method of control variates. In *Monte Carlo and Quasi-Monte Carlo Methods 2000*, pages 27–49. Springer, 2002.

[13] T. Hastie, R. Tibshirani, J. Friedman, and J. Franklin. The elements of statistical learning: data mining, inference and prediction. *The Mathematical Intelligencer*, 27(2):83–85, 2005.
[14] F. J. Hickernell, C. Lemieux, A. B. Owen, et al.Control variates for quasi-Monte Carlo. *Statistical Science*, 20(1):1–31, 2005.
[15] F. Y. Kuo, I. H. Sloan, G. W. Wasilkowski, and B. J. Waterhouse. Randomly shifted lattice rules with the optimal rate of convergence for unbounded integrands. *Journal of Complexity*, 26(2):135–160, 2010.
[16] F. Y. Kuo, W. T. M. Dunsmuir, I. H. Sloan, M. P. Wand, and R. S. Womersley. Quasi-Monte Carlo for highly structured generalised response models. *Methodology and Computing in Applied Probability*, 10(2):239–275, 2008.
[17] S. L. Lohr. *Sampling: Design and Analysis*. Chapman and Hall/CRC, 2019.
[18] A. Mugler and H.-J. Starkloff. On the convergence of the stochastic Galerkin method for random elliptic partial differential equations. *ESAIM: Mathematical Modelling and Numerical Analysis*, 47:1237–1263, 9 2013.
[19] H. Niederreiter. *Random Number Generation and quasi-Monte Carlo Methods*. Society for Industrial and Applied Mathematics, Philadelphia, PA, USA, 1992.
[20] H. Niederreiter. *Random Number Generation and Quasi-Monte Carlo Methods*, volume 63. Siam, 1992.
[21] S. Smolyak. Quadrature and interpolation formulas for tensor products of certain classes of functions. *Soviet Mathematics, Doklady*, 4:240–243, 1963.
[22] X. Wang and I. H. Sloan. Quasi-Monte Carlo methods in financial engineering: An equivalence principle and dimension reduction. *Operations Research*, 59(1):80–95, January 2011.

Michael B. Giles, Mario Hefter, Lukas Mayer, and Klaus Ritter

2 An adaptive random bit multilevel algorithm for SDEs

Abstract: We study the approximation of expectations $\mathrm{E}(f(X))$ for solutions X of stochastic differential equations and functionals f on the path space by means of Monte Carlo algorithms that only use random bits instead of random numbers. We construct an adaptive random bit multilevel algorithm, which is based on the Euler scheme, the Lévy–Ciesielski representation of the Brownian motion, and asymptotically optimal random bit approximations of the standard normal distribution. We numerically compare this algorithm with the adaptive classical multilevel Euler algorithm for a geometric Brownian motion, an Ornstein–Uhlenbeck process, and a Cox–Ingersoll–Ross process.

Keywords: Random bits, multilevel Monte Carlo algorithm, stochastic differential equation, adaption

MSC 2010: 65C30, 65C05, 60H35

2.1 Introduction

We study the approximation of expectations $\mathrm{E}(f(X))$, where $X = (X(t))_{t\in[0,1]}$ is the r-dimensional solution of an autonomous stochastic differential equation (SDE) driven by a d-dimensional Brownian motion, and where $f\colon C([0,1],\mathbb{R}^r) \to \mathbb{R}$ is a functional on the path space. The main contribution of this paper is the construction of an adaptive random bit multilevel algorithm $A_\varepsilon^{\mathrm{bit}}$, which is based on the generic adaptive multilevel algorithm from [5]. Here, $\varepsilon > 0$ is an accuracy demand and input to the algorithm, and the maximal level as well as the replication numbers per level are determined adaptively. For a survey on multilevel Monte Carlo algorithms we refer to [4].

Acknowledgement: Mike Giles was partially supported by the UK Engineering and Physical Science Research Council (EPSRC) through the ICONIC Programme Grant, EP/P020720/1. Lukas Mayer was supported by the Deutsche Forschungsgemeinschaft (DFG) within the RTG 1932 "Stochastic Models for Innovations in the Engineering Sciences".

Michael B. Giles, Mathematical Institute, University of Oxford, Oxford OX2 6GG, England, e-mail: mike.giles@maths.ox.ac.uk
Mario Hefter, Lukas Mayer, Klaus Ritter, Fachbereich Mathematik, Technische Universität Kaiserslautern, Postfach 3049, 67653 Kaiserslautern, Germany, e-mails: hefter@mathematik.uni-kl.de, lmayer@mathematik.uni-kl.de, ritter@mathematik.uni-kl.de

https://doi.org/10.1515/9783110635461-002

The algorithm $A_\varepsilon^{\text{bit}}$ employs the Euler scheme, the Lévy–Ciesielski representation of the Brownian motion (Brownian bridge construction), and the asymptotically optimal random bit approximation of the standard normal distribution according to [7, Thm. 1]. Unfortunately, we have no analysis for the error and the cost of $A_\varepsilon^{\text{bit}}$, and even a proof of the convergence $\lim_{\varepsilon\to 0} A_\varepsilon^{\text{bit}}(f) = \mathrm{E}(f(X))$ (under suitable assumptions on the coefficients of the SDE and the functional f) is missing. Instead we present numerical experiments.

On each level ℓ a multilevel algorithm has to couple a fine approximation and a course approximation. In the classical setting, where random numbers are used, one may simply simulate Brownian increments for $2^\ell + 1$ equidistant points. For the fine approximation the Euler scheme with 2^ℓ steps is applied, and for the course approximation, the step-size is doubled and the corresponding increments are added up. Of course, there are several options for the simulation of the Brownian increments, in particular, one may either simulate the increments directly or use the Lévy–Ciesielski representation of the Brownian motion.

Both of these approaches may be adapted to the random bit setting by approximating all the involved normally distributed random variables by random variables that can be simulated with random bits only. However, in contrast to the classical setting, the two constructions no longer end up in the same distribution. The first approach, where increments are approximated and the approximations are added up, has been presented and analyzed in [6]. Here, the random bit approximations of the increments are independent, but an additional bias is introduced, in contrast to the classical setting.

In this paper, we present the second approach, which has already been sketched in [4, Sec. 10.2], and which employs the Lévy–Ciesielski representation and random bit approximations to its normally distributed coefficients. Here, the approximations of the Brownian increments are no longer independent, which forms an obstacle for an error analysis, but we obtain matching distributions: The distribution of the course approximation on level ℓ coincides with the distribution of the fine approximation on level $\ell - 1$. Furthermore, this approach is well suited as the building block for an adaptive multilevel algorithm.

In the numerical experiments, we apply the adaptive random bit multilevel Euler algorithm $A_\varepsilon^{\text{bit}}$ and the adaptive classical multilevel Euler algorithm A_ε^{c}, which is based on random numbers, for three processes X and functionals f, namely, the maximum of a geometric Brownian motion and the terminal values of an Ornstein–Uhlenbeck process and of a Cox–Ingersoll–Ross process. At first, we compare the building blocks, i. e., the random bit Euler scheme and the classical Euler scheme, in terms of their bias and variance decays. The decays depend on the process X and the functional f under consideration, but in all three cases we observe no essential difference between the random bit and the classical Euler scheme. Next we turn to the adaptive algorithms. In all three cases and for both algorithms, the actual root mean squared error is almost equal to the accuracy demand ε. Finally, to achieve the same root-mean-squared error, the number of random bits needed by $A_\varepsilon^{\mathrm{bit}}$ is only about 4 to 7 times larger than the number of random numbers needed by $A_\varepsilon^{\mathrm{c}}$.

For the terminal value of the Cox–Ingersoll–Ross process, we also apply the truncated Milstein scheme from [9] as the building block, instead of the Euler scheme. This leads to a substantially faster decay of the variance in the classical and in the random bit case. Moreover, the algorithm based on the truncated Milstein scheme and random bits performs as good as the algorithm based on the Euler scheme and random numbers.

This work is partially motivated by reconfigurable architectures, such as field programmable gate arrays (FPGAs). These devices allow users to choose the precision of each individual operation on a bit level and provide a generator for random bits. In the setting and analysis of the present paper, we take into account the latter fact, whereas we ignore all finite precision issues for arithmetic operations. We refer to [2, 11] for the construction and for extensive tests of a finite precision multilevel algorithm for FPGAs with applications in computational finance. For an error analysis of the Euler scheme for SDEs in a finite precision arithmetic, we refer to [12].

2.2 Euler schemes

Consider an autonomous system

$$\mathrm{d}X(t) = a(X(t))\,\mathrm{d}t + b(X(t))\,\mathrm{d}W(t), \quad t \in [0,1],$$

of SDEs with a deterministic initial value $X(0) = x_0 \in \mathbb{R}^r$ and a d-dimensional Brownian motion W, and with drift and diffusion coefficients $a\colon \mathbb{R}^r \to \mathbb{R}^r$ and $b\colon \mathbb{R}^r \to \mathbb{R}^{r\times d}$, respectively. Furthermore, consider the time discretization given by

$$t_{k,\ell} = k/2^\ell, \quad k = 0,\dots,2^\ell, \tag{2.1}$$

for $\ell \in \mathbb{N}_0$ together with a suitable choice of

$$V_\ell = (V_{1,\ell},\dots,V_{2^\ell,\ell}) \tag{2.2}$$

with d-dimensional random vectors $V_{k,\ell}$ on a common probability space. These random vectors are meant to at least approximate the Brownian increments associated with (2.1), and the corresponding Euler scheme X_ℓ is given by $X_\ell(t_{0,\ell}) = x_0$ and

$$X_\ell(t_{k,\ell}) = X_\ell(t_{k-1,\ell}) + 2^{-\ell}\cdot a(X_\ell(t_{k-1,\ell})) + b(X_\ell(t_{k-1,\ell}))\cdot V_{k,\ell} \tag{2.3}$$

for $k = 1,\dots,2^\ell$.

The multilevel approach relies on a coupling of X_ℓ with $\ell \geq 1$ to an Euler scheme $\widetilde{X}_{\ell-1}$ with step-size $2^{-(\ell-1)}$. Hence we choose

$$\widetilde{V}_{\ell-1} = (\widetilde{V}_{1,\ell-1},\dots,\widetilde{V}_{2^{\ell-1},\ell-1}) \tag{2.4}$$

with d-dimensional random vectors $\widetilde{V}_{k,\ell-1}$ on the probability space introduced above, and we define, as before, $\widetilde{X}_{\ell-1}(t_{0,\ell-1}) = x_0$ and

$$\widetilde{X}_{\ell-1}(t_{k,\ell-1}) = \widetilde{X}_{\ell-1}(t_{k-1,\ell-1}) + 2^{-(\ell-1)} \cdot a(\widetilde{X}_{\ell-1}(t_{k-1,\ell-1})) + b(\widetilde{X}_{\ell-1}(t_{k-1,\ell-1})) \cdot \widetilde{V}_{k,\ell-1} \tag{2.5}$$

for $k = 1, \dots, 2^{\ell-1}$. Of course, a natural coupling between X_ℓ and $\widetilde{X}_{\ell-1}$ is induced by

$$\widetilde{V}_{k,\ell-1} = V_{2k,\ell} + V_{2k-1,\ell}, \quad k = 1, \dots, 2^{\ell-1}. \tag{2.6}$$

Actually, the multilevel approach is based on a hierarchy $V_0, (V_1, \widetilde{V}_0), \dots, (V_L, \widetilde{V}_{L-1})$ with corresponding Euler schemes, and the following properties are most convenient for its analysis:

(i) For every $\ell \in \{0, \dots, L\}$, the random vectors $V_{1,\ell}, \dots, V_{2^\ell,\ell}$ are iid with iid real-valued components $V_{k,\ell}^{(1)}, \dots, V_{k,\ell}^{(d)}$.

(ii) For every $\ell \in \{1, \dots, L\}$, the random vectors $\widetilde{V}_{\ell-1}$ and $V_{\ell-1}$ coincide in distribution.

To obtain processes with continuous paths, we extend X_ℓ and $\widetilde{X}_{\ell-1}$ onto $[0, 1]$ by piecewise linear interpolation.

2.2.1 The classical Euler scheme

In the vast majority of papers, $V_{k,\ell}$ and $\widetilde{V}_{k,\ell-1}$ are chosen as Brownian increments, i. e.,

$$V_{k,\ell} = W(t_{k,\ell}) - W(t_{k-1,\ell})$$

and

$$\widetilde{V}_{k,\ell-1} = W(t_{k,\ell-1}) - W(t_{k-1,\ell-1}),$$

so that we have (2.6), and (i) and (ii) are satisfied. Error bounds for the Euler scheme w. r. t. (with respect to) various error criteria and under different sets of assumptions concerning the drift and diffusion coefficients are well known in this case. To simulate the corresponding distributions, a generator for random numbers from $[0, 1]$ has to be available.

2.2.2 Random bit Euler schemes

In the present paper, we study the random bit quadrature problem for SDEs, i. e., we consider algorithms that are only allowed to use random bits instead of random numbers, see also [1, 2, 6, 7, 11, 12]. This excludes the use of Brownian increments.

Heuristics and extensive tests for finite precision random bit multilevel algorithms for field programmable gate arrays (FPGAs) are presented in [2, 11, 12]. In [7] the random bit quadrature problem is studied for Gaussian random fields X, and relations to random bit approximation of Gaussian measures are exploited.

Motivated by the weak error analysis of the Euler scheme, the multilevel construction in [1] is based on iid random vectors $V_{1,L}, \dots, V_{2^L,L}$, each of which has iid components $V_{k,L}^{(1)}, \dots, V_{k,L}^{(d)}$ with

$$2^{L/2-1} \cdot V_{1,L}^{(1)} + 1/2 \sim B(1, 1/2).$$

Moreover, the coupling is defined by (2.6), and (ii) is assumed to hold. It follows that (i) is satisfied as well, and

$$2^{L/2-1} \cdot V_{1,\ell}^{(1)} + 2^{L-\ell-1} \sim B(2^{L-\ell}, 1/2). \tag{2.7}$$

See [1, Sec. 3] for error bounds, and [1, Sec. 4.1] for the discussion of fast generation of random quantities in this context.

A different construction is presented and analyzed in [6]. Here, the starting point is the approximation of the standard normal distribution based on random bits. Let Φ^{-1} denote the inverse of the distribution function of $N(0, 1)$, and let U be uniformly distributed on

$$D_q = \left\{ \sum_{i=1}^{q} b_i \cdot 2^{-i} + 2^{-(q+1)} : b_i \in \{0, 1\} \text{ for } i = 1, \dots, q \right\}, \tag{2.8}$$

where $q \in \mathbb{N}$. Obviously, q random bits suffice to simulate the distribution ν_q of $\Phi^{-1} \circ U$, which serves as an approximation of $N(0, 1)$. Further, properties of ν_q, in particular, error bounds and the weak asymptotic optimality among all approximations based on q random bits, have been established in [7, Sec. 2.2]. In the construction from [6], (i) is assumed to hold with

$$2^{\ell/2} \cdot V_{1,\ell}^{(1)} \sim \nu_L, \tag{2.9}$$

and the coupling is again defined by (2.6). Consequently, the analogon to (i) also holds for the random vectors $\widetilde{V}_{1,\ell-1}, \dots, \widetilde{V}_{2^{\ell-1},\ell-1}$, but property (ii) is not satisfied, which introduces an additional bias term in the multilevel analysis. See [6] for error and cost bounds; in particular, a variant of the corresponding multilevel Euler algorithm, which also employs Bakhvalov's trick, is shown to be almost worst-case optimal on the class of all Lipschitz continuous functionals $f: C([0, 1], \mathbb{R}^r) \to \mathbb{R}$ with Lipschitz constant at most one. Observe that the number of random bits that are needed to simulate the distribution of V_ℓ with $\ell = 0$ or the joint distribution of V_ℓ and $\widetilde{V}_{\ell-1}$ with $\ell \geq 1$ is of the order $d \cdot L \cdot 2^\ell$.

2.2.3 The random bit Lévy–Ciesielski Euler scheme

In the sequel, we present a new construction of a random bit Euler scheme, which is based on the Lévy–Ciesielski representation of the Brownian motion. Hereby, we get matching distributions across the levels in the sense of (ii), but the iid-property (i) is not satisfied. The main advantage of the new construction, compared to the approaches from [1, 6], is that it is well suited as the building block for an adaptive multilevel algorithm.

Consider the sequence of Schauder functions $s^{(i,j)}$ with $i = 0$ and $j = 1$ or $i \in \mathbb{N}$ and $j = 1 \dots, 2^{i-1}$. These functions are given by

$$s^{(i,j)}(t) = \int_0^t h^{(i,j)}(u)\,du, \quad t \in [0,1],$$

with Haar wavelets $h^{(0,1)} = 1$ and

$$h^{(i,j)} = 2^{(i-1)/2} \cdot (1_{I^{(i,j)}} - 1_{J^{(i,j)}})$$

for $i \in \mathbb{N}$ and $j = 1, \dots, 2^{i-1}$, where

$$I^{(i,j)} = [(j-1)/2^{i-1}, (j-1/2)/2^{i-1}[$$

and

$$J^{(i,j)} = [(j-1/2)/2^{i-1}, j/2^{i-1}[.$$

The Lévy–Ciesielski representation states that

$$W_\ell = s^{(0,1)} \cdot Z^{(0,1)} + \sum_{i=1}^{\ell} \sum_{j=1}^{2^{i-1}} s^{(i,j)} \cdot Z^{(i,j)} \tag{2.10}$$

with an independent family of d-dimensional standard normally distributed random vectors $Z^{(i,j)}$ with $i = 0$ and $j = 1$, or $i \in \mathbb{N}$ and $j = 1 \dots, 2^{i-1}$ converges to a d-dimensional Brownian motion as $\ell \to \infty$, e. g., in mean square and almost surely w. r. t. the supremum-norm. We add that

$$W_n(t_{k,\ell}) = W_\ell(t_{k,\ell}), \quad k = 0, \dots, 2^\ell, \tag{2.11}$$

for $\ell, n \in \mathbb{N}_0$ with $\ell < n$. In this sense, W_ℓ already yields the values of the Brownian motion at the discretization (2.1).

In a random bit approximation that corresponds to W_ℓ, the number of bits that are spent for the individual terms should depend on i and ℓ, but not on the shift parameter j. We spend

$$q_\ell^{(i)} = 2 \cdot (\ell + 1 - i)$$

random bits for the approximation of the distribution of each of the components of $Z^{(i,j)}$. This choice is motivated by [7, Thm. 2], which determines the weak asymptotics for random bit approximation of a Brownian bridge with respect to the L_2-norm.

Accordingly, we consider an independent sequence $U_\ell^{(0,1)}, \dots, U_\ell^{(\ell,2^{\ell-1})}$ of d-dimensional random vectors, with iid components, that are uniformly distributed on

$$D_\ell^{(i)} = D_{q_\ell^{(i)}}.$$

To normalize the variances, we put

$$\sigma_\ell^{(i)} = 2^{-q_\ell^{(i)}/2} \cdot \Big(\sum_{x \in D_\ell^{(i)}} (\Phi^{-1}(x))^2 \Big)^{1/2}.$$

Replacing $Z^{(i,j)}$ by

$$Y_\ell^{(i,j)} = 1/\sigma_\ell^{(i)} \cdot \Phi^{-1} \circ U_\ell^{(i,j)}$$

in (2.10), where $\Phi^{-1} \circ U_\ell^{(i,j)}$ denotes the application of Φ^{-1} to every component of $U_\ell^{(i,j)}$, we obtain a random bit counterpart to W_ℓ.

Next, we turn to the Brownian increments, and we put

$$\Delta_{k,\ell}^{(i,j)} = s^{(i,j)}(t_{k,\ell}) - s^{(i,j)}(t_{k-1,\ell}).$$

We use (2.2) with

$$V_{k,\ell} = \Delta_{k,\ell}^{(0,1)} \cdot Y_\ell^{(0,1)} + \sum_{i=1}^{\ell} \sum_{j=1}^{2^{i-1}} \Delta_{k,\ell}^{(i,j)} \cdot Y_\ell^{(i,j)}$$

to approximate, in distribution, the Brownian increments corresponding to (2.1).

Lemma 2.1. *The components of $V_{k,\ell}$ have mean zero and variance $2^{-\ell}$.*

The normalization is crucial in the definition of the random vectors $Y_\ell^{(i,j)}$. In fact, without this normalization the variances of the Brownian increments are not even matched asymptotically, and thus one cannot expect the Euler scheme to converge, in any reasonable sense, to the true solution of the SDE.

Lemma 2.2. *The components of*

$$V'_{k,\ell} = \Delta_{k,\ell}^{(0,1)} \cdot \Phi^{-1} \circ U_\ell^{(0,1)} + \sum_{i=1}^{\ell} \sum_{j=1}^{2^{i-1}} \Delta_{k,\ell}^{(i,j)} \cdot \Phi^{-1} \circ U_\ell^{(i,j)}$$

have mean zero and variance at most $0.9 \cdot 2^{-\ell}$ for $\ell \geq 1$.

See the Appendix for the proofs of Lemma 2.1 and Lemma 2.2.

Let $\ell \geq 1$. For the multilevel construction, we have to couple V_ℓ in a suitable way to a random vector $\widetilde{V}_{\ell-1}$ that approximates, in distribution, the Brownian increments with step-size $2^{-(\ell-1)}$. To this end, we introduce the rounding function

$$T_q\colon [0,1[\to D_q, \quad x \mapsto \frac{\lfloor 2^q x \rfloor}{2^q} + 2^{-(q+1)},$$

see (2.8), and we put

$$T_{\ell-1}^{(i)} = T_{q_{\ell-1}^{(i)}}$$

for $i = 0, \dots, \ell - 1$ to obtain

$$U_{\ell-1}^{(i,j)} \stackrel{\mathrm{d}}{=} T_{\ell-1}^{(i)} \circ U_\ell^{(i,j)} \tag{2.12}$$

for $j = 1$ if $i = 0$, and for $j = 1, \dots, 2^{i-1}$ if $i \geq 1$. We define

$$\widetilde{Y}_{\ell-1}^{(i,j)} = 1/\sigma_{\ell-1}^{(i)} \cdot \Phi^{-1} \circ T_{\ell-1}^{(i)} \circ U_\ell^{(i,j)},$$

and we use (2.4) with

$$\widetilde{V}_{k,\ell-1} = \Delta_{k,\ell-1}^{(0,1)} \cdot \widetilde{Y}_{\ell-1}^{(0,1)} + \sum_{i=1}^{\ell-1} \sum_{j=1}^{2^{i-1}} \Delta_{k,\ell-1}^{(i,j)} \cdot \widetilde{Y}_{\ell-1}^{(i,j)}.$$

Observe that $q_\ell^{(i)} - q_{\ell-1}^{(i)} = 2$. Hence, $\widetilde{V}_{\ell-1}$ is, roughly speaking, obtained from V_ℓ by ignoring the two least important bits in all of the relevant terms.

We stress that neither V_ℓ nor $\widetilde{V}_{\ell-1}$ has independent components, except for the trivial cases $\ell = 0$ or $\ell = 1$, respectively, so that (i) is not satisfied. On the other hand, (2.12) implies that we have matching distributions in the sense of (ii).

The number of random bits that are needed to simulate the distribution of V_ℓ with $\ell = 0$ or the joint distribution of V_ℓ and $\widetilde{V}_{\ell-1}$ with $\ell \geq 1$ is given by

$$d \cdot \left(q_\ell^{(0)} + \sum_{i=1}^{\ell} 2^{i-1} \cdot q_\ell^{(i)} \right) = d \cdot (2^{\ell+2} - 2),$$

which is easily verified by induction, cf. [7, Thm. 2]. Furthermore, the arithmetic cost to compute V_ℓ, together with $\widetilde{V}_{\ell-1}$ if $\ell \geq 1$, is of the order $d \cdot 2^\ell$, see, e. g., [10, Sec. 2.2].

Let us discuss two important differences between the two constructions from [1, 6], which have been discussed in Section 2.2.2, and the construction based on the Lévy-Ciesielski representation.

The maximal level L has to be known in advance for the former two constructions, see (2.7) and (2.9), whereas this is not the case for the latter construction. Due to this difference, the Lévy-Ciesielski-based construction is well suited as the building block for an adaptive multilevel algorithm.

On the other hand, analytic results are only available for the constructions from [1, 6], since we have (i) and its analogon for $\widetilde{V}_{\ell-1}$ only in these two cases.

2.3 Adaptive algorithms and experiments

We consider an adaptive multilevel algorithm with either one of the following building blocks:
1. The random bit Euler schemes with V_ℓ and $\widetilde{V}_\ell$ based on the Lévy–Ciesielski representation, see Section 2.2.3. Notation: X_ℓ^{bit} and $\widetilde{X}_{\ell-1}^{\mathrm{bit}}$.
2. The classical Euler schemes with V_ℓ and $\widetilde{V}_\ell$ based on the Brownian increments, see Section 2.2.1. Notation: X_ℓ^{c} and $\widetilde{X}_{\ell-1}^{\mathrm{c}}$.

The number of calls to the random number generator as well as the number of arithmetic operations to jointly simulate X_ℓ^* and $\widetilde{X}_{\ell-1}^*$ with $\ell \geq 1$ is of the order 2^ℓ for both variants.

In both cases, we use the adaptive algorithm A_ε^* from [5]. Here, $\varepsilon > 0$ is an accuracy demand and input to the algorithm, and the maximal level as well as the replication numbers per level are determined adaptively.

We present numerical results for three different scalar SDEs, i. e., $r = d = 1$, where the solutions $\mathrm{E}(f(X))$ are known analytically. For a fixed SDE and a fixed functional f, we put

$$\mathrm{bias}_\ell^* = \mathrm{E}(f(X_\ell^*) - f(X_{\ell-1}^*))$$

and

$$\mathrm{var}_\ell^* = \mathrm{Var}(f(X_\ell^*) - f(X_{\ell-1}^*))$$

for $\ell \geq 1$. As key quantities, we consider the root mean squared error

$$\mathrm{rmse}_\varepsilon^* = \left(\mathrm{E}(A_\varepsilon^*(f) - \mathrm{E}(f(X)))^2\right)^{1/2}$$

of A_ε^*, applied to f for the particular SDE, and the corresponding cost

$$\mathrm{cost}_\varepsilon^* = \mathrm{E}(C_\varepsilon^*(f)),$$

where $C_\varepsilon^*(f)$ denotes the number of calls of the random number generator for $\varepsilon > 0$. All of these quantities can be approximated by simple Monte Carlo algorithms, and the corresponding results will be presented together with asymptotic confidence intervals with confidence level 0.95 in the sequel. The number of Monte Carlo replications for the data points and confidence intervals, involving root mean squared errors, varies between $2 \cdot 10^3$ and $2 \cdot 10^4$.

2.3.1 Geometric Brownian motion

Here, we consider the geometric Brownian motion X that solves

$$\mathrm{d}X(t) = 1/50 \cdot X(t)\,\mathrm{d}t + 1/5 \cdot X(t)\,\mathrm{d}W(t), \quad t \in [0, 1],$$

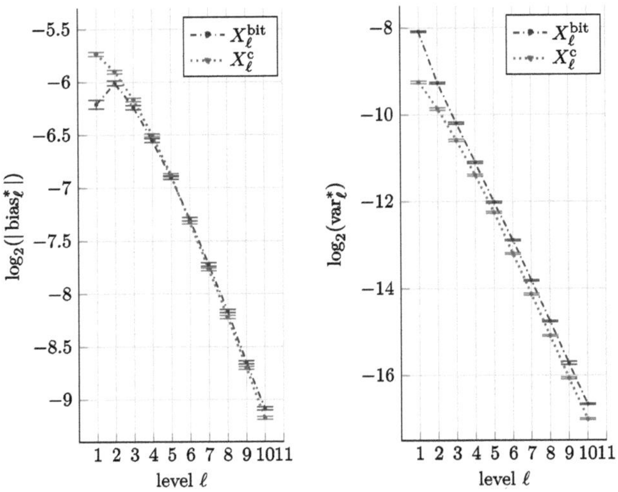

Figure 2.1: Maximum of a geometric Brownian motion: bias and variance vs. level.

with initial value $x_0 = 1$, as well as the path-dependent functional given by

$$f(x) = \max_{0 \le t \le 1} x(t).$$

Since $X(t) = \exp(W(t)/5)$, we obtain

$$\mathrm{E}(f(X)) = \mathrm{E}(\exp(|W(1)|/5)) = (2/\pi)^{1/2} \cdot \int_0^\infty \exp(y/5 - y^2/2)\,dy = 1.1819\ldots.$$

At first, we compare the random bit Euler scheme X_ℓ^{bit} and the classical Euler scheme X_ℓ^{c} in terms of their bias and variance, see Figure 2.1. Since f is Lipschitz continuous w. r. t. the supremum norm on $C([0,1])$, we have the well-known upper bound

$$\mathrm{var}_\ell^{\mathrm{c}} = O(\ell \cdot 2^{-\ell})$$

and, consequently,

$$|\mathrm{bias}_\ell^{\mathrm{c}}| = O(\ell^{1/2} \cdot 2^{-\ell/2}).$$

These upper bounds are very well reflected in the actual bias and variance decays, and we observe no essential difference between the random bit and the classical Euler scheme.

Next, we compare the multilevel algorithms $A_\varepsilon^{\mathrm{bit}}$ and $A_\varepsilon^{\mathrm{c}}$. At first, we relate the root mean squared error $\mathrm{rmse}_\varepsilon^*$ to the accuracy demand ε, see Figure 2.2, where we consider 25 different values of ε in a reasonable range. For both algorithms the root mean squared error is almost equal to the accuracy demand.

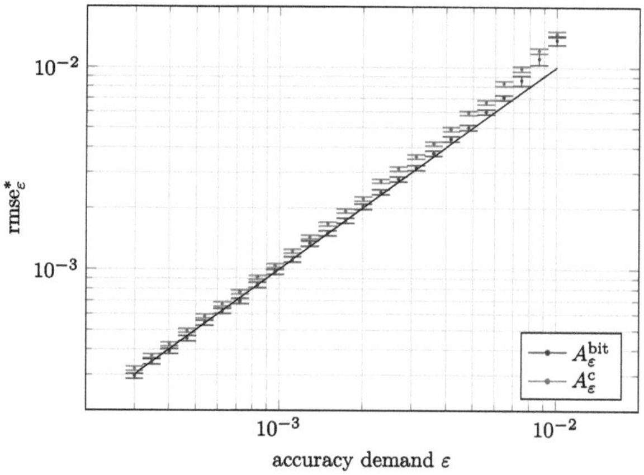

Figure 2.2: Maximum of a geometric Brownian motion: root mean squared error vs. accuracy demand.

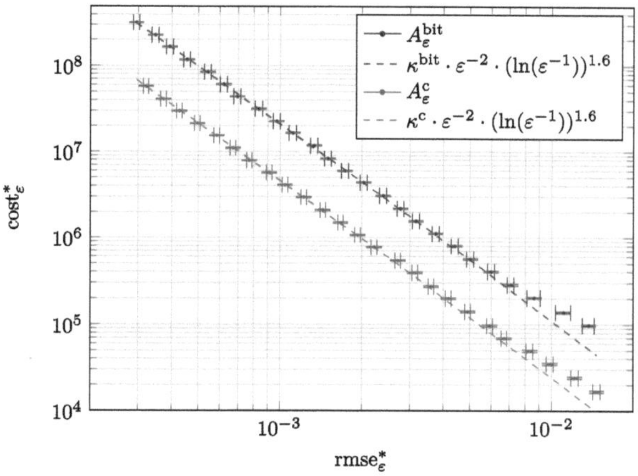

Figure 2.3: Maximum of a geometric Brownian motion: cost vs. root mean squared error.

Finally we relate $\mathrm{cost}^*_\varepsilon$ to the root mean squared error $\mathrm{rmse}^*_\varepsilon$, see Figure 2.3, which is based on the same data set as Figure 2.2. We add that the confidence intervals for $\mathrm{cost}^*_\varepsilon$ in Figure 2.3 are rather small and hardly visible. Figure 2.3 includes two graphs of functions $\varepsilon \mapsto \kappa \cdot \varepsilon^{-2} \cdot (\ln(\varepsilon^{-1}))^\gamma$ with parameters $\kappa > 0$ and $\gamma \in \mathbb{R}$, which are fitted to the respective data by hand. We obtain a log-exponent of $\gamma = 1.6$ as a good fit for both algorithms. The presence of a logarithmic term, i. e., $\gamma \neq 0$, corresponds to the actual bias and variance decays. The number of random bits is roughly $\kappa^{\mathrm{bit}}/\kappa^{\mathrm{c}} = 4.57$ times larger than the number of random numbers for the same root mean squared error.

2.3.2 Ornstein–Uhlenbeck process

Here, we consider the Ornstein–Uhlenbeck process X that solves

$$dX(t) = (2 - X(t))\,dt + dW(t), \quad t \in [0,1],$$

with initial value $x_0 = 1$, as well as the path-independent functional given by

$$f(x) = x(1). \tag{2.13}$$

Since $X(t) = \exp(-t) + 2(1 - \exp(-t)) + \int_0^t \exp(-(t-s))\,dW_s$, we obtain

$$E(f(X)) = 2 - \exp(-1) = 1.6321\ldots.$$

As a major difference to the previous example, we have improved upper bounds for the classical Euler scheme X_ℓ^c, namely,

$$\mathrm{var}_\ell^c = O(2^{-2\ell}) \tag{2.14}$$

and, consequently,

$$|\mathrm{bias}_\ell^c| = O(2^{-\ell}). \tag{2.15}$$

For the numerical experiments, we proceed as in the previous section. As for the geometric Brownian motion, the upper bounds (2.14) and (2.15) are very well reflected in the actual bias and variance decays, and we observe no essential difference between the random bit and the classical Euler scheme, see Figure 2.4.

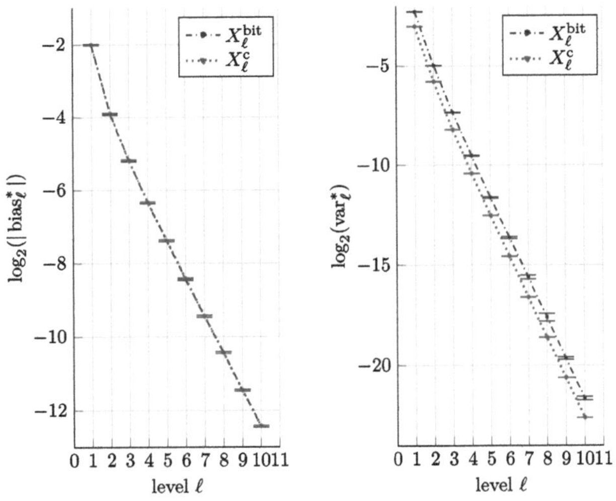

Figure 2.4: Ornstein–Uhlenbeck process at final time: bias and variance vs. level.

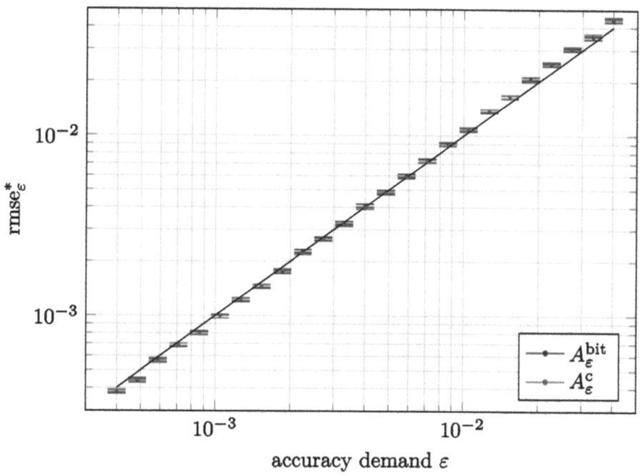

Figure 2.5: Ornstein–Uhlenbeck process at final time: root-mean-squared error vs. accuracy demand.

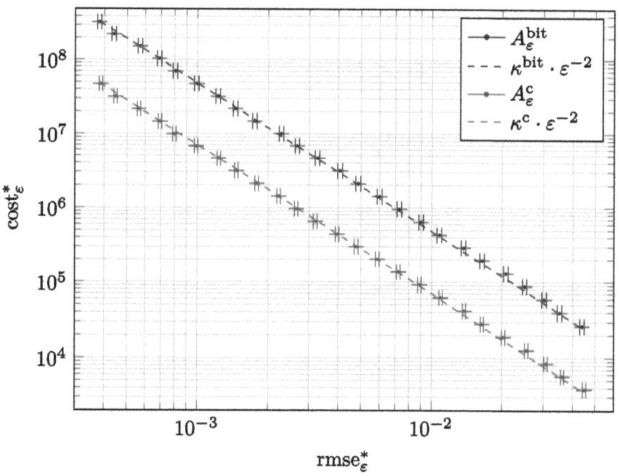

Figure 2.6: Ornstein–Uhlenbeck process at final time: cost vs. root-mean-squared error.

For both multilevel algorithms, the root-mean-squared error is again almost equal to the accuracy demand, see Figure 2.5.

Due to the improved upper bounds for the variance and bias, it is natural to expect that $\text{cost}^*_\varepsilon$ is proportional to $(\text{rmse}^*_\varepsilon)^{-2}$. This is in line with the numerical results in Figure 2.6. Furthermore, we have $\kappa^{\text{bit}}/\kappa^{\text{c}} = 6.85$.

2.3.3 Cox–Ingersoll–Ross Process

Here, we consider the Cox–Ingersoll–Ross process X that solves

$$\mathrm{d}X(t) = (3/2 - X(t))\,\mathrm{d}t + 2\cdot\sqrt{X(t)}\,\mathrm{d}W(t), \quad t \in [0,1],$$

with initial value $x_0 = 1$, as well as f given by (2.13). We have

$$\mathrm{E}(f(X)) = \exp(-1) + \frac{3}{2}(1 - \exp(-1)) = 1.3160\ldots,$$

see, e. g., [3, eqn. (19)]. To get a well-defined variant of the Euler scheme, we take the positive part in every Euler step, i. e., we take the maximum with 0 of the right-hand side in (2.3) and (2.5).

Furthermore, we compare this Euler scheme with a truncated Milstein scheme, which is proposed and analyzed in [9]. For this scheme, the right-hand side of (2.3), and similarly also for (2.5), is replaced by $\Theta_{\mathrm{tMil}}(X_\ell(t_{k-1,\ell}), 2^{-\ell}, V_{k,\ell})$, where

$$\Theta_{\mathrm{tMil}}(x, h, w) = \max(0, (\max(\sqrt{h}, \sqrt{\max(h, x)} + w))^2 + (1/2 - x) \cdot h).$$

The resulting schemes are denoted by $X_\ell^{\mathrm{c,tMil}}$ and $X_\ell^{\mathrm{bit,tMil}}$. For the Euler scheme, no polynomial strong convergence rate is known. For the truncated Milstein scheme, the strong convergence that result from [9, Thm. 1] implies

$$\mathrm{var}_\ell^{\mathrm{c,tMil}} = O(2^{-\ell/2+\varepsilon\cdot\ell}) \tag{2.16}$$

and

$$|\mathrm{bias}_\ell^{\mathrm{c,tMil}}| = O(2^{-\ell/2+\varepsilon\cdot\ell}) \tag{2.17}$$

for every $\varepsilon > 0$. This strong convergence rate is the best known convergence rate for the Cox–Ingersoll–Ross process, see [8, Fig. 1.1].

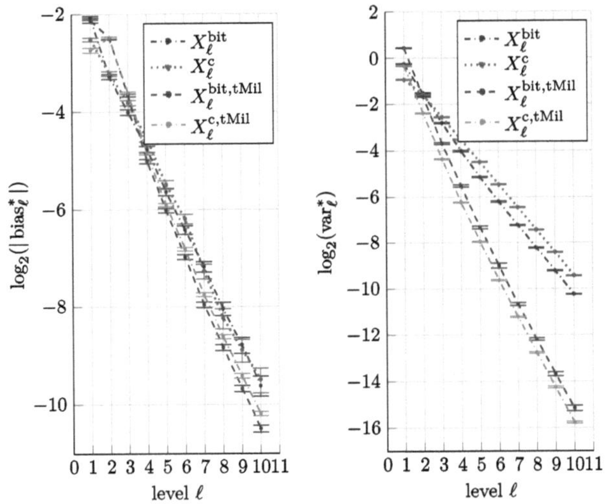

Figure 2.7: Cox–Ingersoll–Ross process at final time: bias and variance vs. level.

For the numerical experiments, we proceed as in the previous sections. At first, we comment on Figure 2.7. The decay of the bias for all four variants is similar to the decay of the bias for the Ornstein–Uhlenbeck process. The decay of the variance for both variants, based on the Euler scheme, is similar to the decay of the variance for the geometric Brownian motion. The decay of the variance for both versions, based on the truncated Milstein scheme, is similar and substantially faster. Note that the upper bounds (2.16) and (2.17) seem to be too pessimistic, cf. the conjecture in [9, Fig. 5].

For all four algorithms, the root-mean-squared error is almost equal to the accuracy demand, see Figure 2.8, as is the case of the SDEs considered before.

Finally, we relate $\mathrm{cost}_\varepsilon^*$ to the root-mean-squared error $\mathrm{rmse}_\varepsilon^*$, see Figure 2.9. The exponent γ of the logarithmic term is equal to 1.2 for both variants that are based on the Euler scheme, and equal to 0.5 for both variants that are based on the truncated

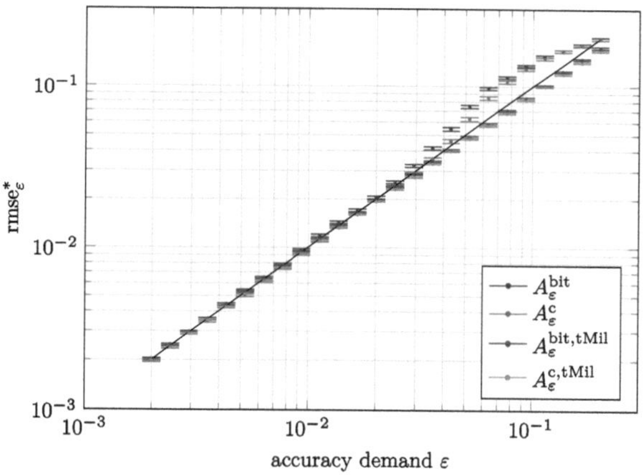

Figure 2.8: Cox–Ingersoll–Ross process at final time: root-mean-squared error vs. accuracy demand.

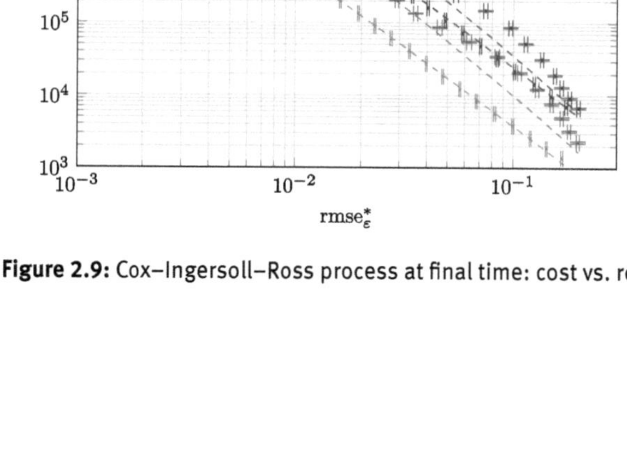

Figure 2.9: Cox–Ingersoll–Ross process at final time: cost vs. root-mean-squared error.

Milstein scheme. The better log-exponent corresponds to the faster decay of the variances. Furthermore, we have $\kappa^{\mathrm{bit}}/\kappa^{\mathrm{c}} = 3.51$ and $\kappa^{\mathrm{bit,tMil}}/\kappa^{\mathrm{c,tMil}} = 6.52$, and the multilevel algorithm based on the Euler scheme with random numbers has roughly the same root-mean-squared error as the multilevel algorithm based on the truncated Milstein scheme using random bits in the range considered in Figure 2.9.

Appendix

We present the proofs of Lemma 2.1 and Lemma 2.2. For notational convenience, we consider the case $d = 1$. Recall that $Z^{(0,1)}, \dots, Z^{(\ell,2^{\ell-1})}$ are independent and standard normally distributed.

The distribution of $U_\ell^{(i,j)}$ is symmetric with respect to 1/2, so that $\mathrm{E}(Y_\ell^{(i,j)}) = 0$. Moreover, $\mathrm{Var}(Y_\ell^{(i,j)}) = 1$ by construction. It follows that $\mathrm{E}(V_{k,\ell}) = 0$ and

$$\begin{aligned}\mathrm{Var}(V_{k,\ell}) &= (\Delta_{k,\ell}^{(0,1)})^2 + \sum_{i=1}^{\ell}\sum_{j=1}^{2^{i-1}} (\Delta_{k,\ell}^{(i,j)})^2 \\ &= \mathrm{Var}\left(\Delta_{k,\ell}^{(0,1)} \cdot Z^{(0,1)} + \sum_{i=1}^{\ell}\sum_{j=1}^{2^{i-1}} \Delta_{k,\ell}^{(i,j)} \cdot Z^{(i,j)}\right) \\ &= \mathrm{Var}(W_\ell(t_{k,\ell}) - W_\ell(t_{k-1,\ell})) = 2^{-\ell}\end{aligned}$$

due to (2.11) and the convergence of the Lévy-Ciesielski representation.

We have

$$\mathrm{Var}(\Phi^{-1} \circ U_\ell^{(i,j)}) \le 1,$$

see the end of the proof of [7, Thm. 1], and in particular for $i = \ell$

$$\mathrm{Var}(\Phi^{-1} \circ U_\ell^{(\ell,j)}) \le 4/5,$$

which follows from a simple computation. It follows that

$$\begin{aligned}\mathrm{Var}(V'_{k,\ell}) &= (\Delta_{k,\ell}^{(0,1)})^2 \cdot \mathrm{Var}(\Phi^{-1} \circ U_\ell^{(0,1)}) + \sum_{i=1}^{\ell}\sum_{j=1}^{2^{i-1}} (\Delta_{k,\ell}^{(i,j)})^2 \cdot \mathrm{Var}(\Phi^{-1} \circ U_\ell^{(i,j)}) \\ &\le (\Delta_{k,\ell}^{(0,1)})^2 + \sum_{i=1}^{\ell-1}\sum_{j=1}^{2^{i-1}} (\Delta_{k,\ell}^{(i,j)})^2 + \sum_{j=1}^{2^{\ell-1}} (\Delta_{k,\ell}^{(\ell,j)})^2 \cdot \mathrm{Var}(\Phi^{-1} \circ U_\ell^{(\ell,j)}) \\ &\le 2^{-\ell} - \sum_{j=1}^{2^{\ell-1}} (\Delta_{k,\ell}^{(\ell,j)})^2 \cdot (1 - \mathrm{Var}(\Phi^{-1} \circ U_\ell^{(\ell,j)})) \\ &\le 2^{-\ell} - 1/5 \cdot \sum_{j=1}^{2^{\ell-1}} (\Delta_{k,\ell}^{(\ell,j)})^2 = 9/10 \cdot 2^{-\ell}.\end{aligned}$$

Bibliography

[1] D. Belomestny and T. Nagapetyan. Multilevel path simulation for weak approximation schemes with application to Lévy-driven SDEs. *Bernoulli*, 23(2):927–950, 2017.

[2] C. Brugger, C. De Schryver, N. Wehn, S. Omland, M. Hefter, K. Ritter, A. Kostiuk, and R. Korn. Mixed precision multilevel Monte Carlo on hybrid computing systems. In *2014 IEEE Conference on Computational Intelligence for Financial Engineering Economics (CIFEr)*, pages 215–222, March 2014.

[3] J. C. Cox, J. E. Ingersoll, Jr., and S. A. Ross. A theory of the term structure of interest rates. *Econometrica*, 53(2):385–407, 1985.

[4] M. B. Giles. Multilevel Monte Carlo methods. *Acta Numer.*, 24:259–328, 2015.

[5] M. B. Giles. Multilevel Monte Carlo software. http://people.maths.ox.ac.uk/~gilesm/mlmc, accessed: Jan. 2019.

[6] M. B. Giles, M. Hefter, L. Mayer, and K. Ritter. Random bit multilevel algorithms for stochastic differential equations. *J. Complex.*, 54:101395, 2019.

[7] M. B. Giles, M. Hefter, L. Mayer, and K. Ritter. Random bit quadrature and approximation of distributions on Hilbert spaces. *Found. Comput. Math.*, 19(1):205–238, 2019.

[8] M. Hefter and A. Herzwurm. Optimal strong approximation of the one-dimensional squared Bessel process. *Commun. Math. Sci.*, 15(8):2121–2141, 2017.

[9] M. Hefter and A. Herzwurm. Strong convergence rates for Cox-Ingersoll-Ross processes—Full parameter range. *J. Math. Anal. Appl.*, 459(2):1079–1101, 2018.

[10] G. Leobacher. Fast orthogonal transforms and generation of Brownian paths. *J. Complex.*, 28(2):278–302, 2012.

[11] S. Omland, M. Hefter, K. Ritter, C. Brugger, C. De Schryver, N. Wehn, and A. Kostiuk. Exploiting mixed-precision arithmetics in a multilevel Monte Carlo approach on FPGAs. In C. De Schryver, editor, *FPGA Based Accelerators for Financial Applications*, pages 191–220. Springer, 2015.

[12] S. Omland. *Mixed Precision Multilevel Monte Carlo Algorithms for Reconfigurable Hardware Systems*. PhD thesis, Technische Universität Kaiserslautern, 2016.

Kerstin Hesse

3 RBF-based penalized least-squares approximation of noisy scattered data on the sphere

Abstract: This paper reports on some recent developments in global and local RBF-based penalized least-squares approximation on the sphere from noisy scattered data. Given is noisy scattered data $f(\boldsymbol{x}_i)$, $i = 1, 2, \ldots, N$, at N distinct points $\boldsymbol{x}_1, \boldsymbol{x}_2, \ldots, \boldsymbol{x}_N \in \overline{\Omega}$ from a function f in the Sobolev (Hilbert) space $W_2^s(\Omega)$, $s > d/2$, where Ω is either the whole unit sphere $\mathbb{S}^d$ in $\mathbb{R}^{d+1}$ or a nonempty connected open subset of $\mathbb{S}^d$ with Lipschitz-continuous boundary, contained in an open hemisphere. The penalized least-squares approximation of f is defined as the unique minimizer in $W_2^s(\mathbb{S}^d)$ of a Tikhonov-type functional that is the sum of a term that measures the squared norm of the ℓ_2 discrepancy of the approximation from the noisy scattered data, and of the square of a suitably chosen norm for $W_2^s(\mathbb{S}^d)$ multiplied by the regularization parameter $\lambda > 0$. The regularization parameter λ balances between data fitting and the smoothness of the approximation. Choosing the norm for $W_2^s(\mathbb{S}^d)$ in the minimized functional as the native space norm of a radial basis function (RBF), whose native space is $W_2^s(\mathbb{S}^d)$, the penalized least-squares approximation is an RBF approximation computed by solving a linear system with a positive definite symmetric matrix. The regularization parameter $\lambda > 0$ is chosen with Morozov's discrepancy principle, an a posteriori parameter choice strategy. The reported recent results are error estimates for the $L_2(\Omega)$ approximation error that are order optimal for noisy scattered data on well-distributed point-sets $\widetilde{X}_N = \{\boldsymbol{x}_1, \boldsymbol{x}_2, \ldots, \boldsymbol{x}_N\} \subseteq \overline{\Omega}$.

Keywords: Global and local L_2 error estimates, Morozov's discrepancy principle, penalized least-squares approximation, radial basis functions, smoothing approximation, Sobolev spaces, sphere

MSC 2010: 41A15, 65D07, 65D10, 33C55, 42C10

Acknowledgement: The author gratefully acknowledges the support of the RICAM special fall 2018 semester on "Multivariate Algorithms and their Foundations in Number Theory," during which the results from [10, 11] were presented at the workshop on "Multivariate Algorithms and Information-Based Complexity" (5–9 November 2018) at the Johann Radon Institute for Computational and Applied Mathematics (RICAM) in Linz, Austria.

Kerstin Hesse, Institute of Mathematics, Paderborn University, 33098 Paderborn, Germany, Tel.: +49 5251 602605; Fax: +49 5251 603836; e-mail: kerstin.hesse@math.upb.de, https://orcid.org/0000-0003-4125-1941

https://doi.org/10.1515/9783110635461-003

3.1 Introduction

The approximation of global or local noisy scattered data on the 2-dimensional unit sphere as the most basic model of the earth's surface plays an important role in many geophysical applications (see e. g. [4–9]) and is therefore of great interest, both from the mathematical point of view and for application problems.

This paper reviews recent results from [10, 11] on global and local RBF-based penalized least-squares approximation on the sphere from (global or local) noisy scattered data. Discussing the main ideas from [10, 11], without going into the rather technical details, necessitates that some of the definitions and ideas will be given intuitively rather than formally. For all the technical details, the reader is referred to [10, 11].

In this paper, let $d \in \mathbb{N}$ satisfy $d \geq 2$. Let $\boldsymbol{x} \cdot \boldsymbol{y}$, where $\boldsymbol{x}, \boldsymbol{y} \in \mathbb{R}^{d+1}$, denote the Euclidean inner product, and let $\|\boldsymbol{x}\|_2 := (\sum_{j=1}^{d+1} x_j^2)^{1/2}$ denote the induced Euclidean norm on $\mathbb{R}^{d+1}$. Let Ω be either the *(unit) sphere*

$$\mathbb{S}^d := \{\boldsymbol{x} \in \mathbb{R}^{d+1} \ : \ \|\boldsymbol{x}\|_2 = 1\}$$

of $\mathbb{R}^{d+1}$ or a *nonempty connected open subset of* $\mathbb{S}^d$ *with Lipschitz-continuous boundary*, contained in an open hemisphere. (The Lipschitz-continuity of $\partial\Omega$ means that, after possibly rotating the sphere and then mapping Ω with a suitable stereographic projection into $\mathbb{R}^d$, the image of Ω has Lipschitz-continuous boundary in $\mathbb{R}^d$.) For a point-set $X_N = \{\boldsymbol{x}_1, \boldsymbol{x}_2, \ldots, \boldsymbol{x}_N\} \subseteq \overline{\Omega} \subseteq \mathbb{S}^d$ of N points, the *mesh norm of* X_N *with respect to* Ω is defined by

$$h_{X_N,\overline{\Omega}} := \sup_{\boldsymbol{y}\in\overline{\Omega}} \min_{i=1,\ldots,N} \operatorname{dist}(\boldsymbol{y}, \boldsymbol{x}_i),$$

where $\operatorname{dist}(\boldsymbol{x}, \boldsymbol{y}) := \arccos(\boldsymbol{x} \cdot \boldsymbol{y})$ is the geodesic distance between $\boldsymbol{x}$ and $\boldsymbol{y}$.

We are interested in the following approximation problem: Given *noisy scattered data*

$$f(\boldsymbol{x}_i) + \epsilon_i, \quad i = 1, 2, \ldots, N, \tag{3.1}$$

of a function f in $W_2^s(\Omega)$, with $s > d/2$, on a point-set $X_N = \{\boldsymbol{x}_1, \boldsymbol{x}_2, \ldots, \boldsymbol{x}_N\} \subseteq \overline{\Omega}$ of N distinct points, we seek a (with respect to the $L_2(\Omega)$ norm) good approximation of this function. The space $W_2^s(\Omega)$, equipped with the norm $\|\cdot\|_{W_2^s(\Omega)}$, is the *Sobolev (Hilbert) space of order s*, defined via an atlas and an associated partition of unity, considering the unit sphere as a d-dimensional differentiable submanifold of $\mathbb{R}^{d+1}$. For $s \in \mathbb{N}_0$, the space $W_2^s(\Omega)$ contains those functions in $L_2(\Omega)$ whose generalized distributional derivatives up to order s are in $L_2(\Omega)$. The assumption $s > d/2$ guarantees that each equivalence class of functions in $W_2^s(\Omega)$ can be represented by a uniquely determined continuous function on $\overline{\Omega}$. Hence, for $s > d/2$, $W_2^s(\Omega)$ is embedded into the space $C(\overline{\Omega})$ of continuous functions on $\overline{\Omega}$. – The noise $\epsilon_1, \epsilon_2, \ldots, \epsilon_N$ in the data (3.1) is assumed to be *deterministic*, that is, we do not assume a probability distribution for the noise.

3.2 Radial basis function approximation

A *radial basis function (RBF)* for $\mathbb{S}^d$ is a continuous, symmetric, *strictly positive definite*, and *zonal* function $\phi : \mathbb{S}^d \times \mathbb{S}^d \to \mathbb{R}$, that is, ϕ satisfies the following: (i) ϕ is continuous; (ii) $\phi(\boldsymbol{x},\boldsymbol{y}) = \phi(\boldsymbol{y},\boldsymbol{x})$ for all $\boldsymbol{x},\boldsymbol{y} \in \mathbb{S}^d$ (ϕ is symmetric); (iii) the symmetric matrix $[\phi(\boldsymbol{x}_i,\boldsymbol{x}_j)]_{i,j=1,2,\dots,N} \in \mathbb{R}^{N\times N}$ is positive definite for any set $X_N = \{\boldsymbol{x}_1,\boldsymbol{x}_2,\dots,\boldsymbol{x}_N\}$ of N distinct points on $\mathbb{S}^d$ and for any $N \in \mathbb{N}$ (ϕ is strictly positive definite); and (iv) $\phi(\boldsymbol{x},\boldsymbol{y})$ does only depend on the Euclidean inner product $\boldsymbol{x}\cdot\boldsymbol{y}$ or equivalently on the geodesic distance $\operatorname{dist}(\boldsymbol{x},\boldsymbol{y})$ (ϕ is zonal).

Since an RBF ϕ is zonal, the function $\boldsymbol{y} \mapsto \phi(\boldsymbol{y},\boldsymbol{x})$, with fixed $\boldsymbol{x} \in \mathbb{S}^d$, of one variable on $\mathbb{S}^d$ is rotationally symmetric about the axis through the origin and $\boldsymbol{x}$. Radial basis functions may have local support (see, e. g., [18, Chapter 9]).

In *RBF approximation on* $\mathbb{S}^d$, we seek an approximation of the form

$$\sum_{j=1}^{N} \alpha_j\, \phi(\cdot,\boldsymbol{x}_j),$$

that is, for a point-set $X_N = \{\boldsymbol{x}_1,\boldsymbol{x}_2,\dots,\boldsymbol{x}_N\} \subseteq \mathbb{S}^d$ of N distinct points, we consider all linear combinations of the (rotationally symmetric) shifted copies $\phi(\cdot,\boldsymbol{x}_j)$, $j = 1,2,\dots,N$, of $\phi(\cdot,\boldsymbol{x})$. In our case, the point-set $X_N = \{\boldsymbol{x}_1,\boldsymbol{x}_2,\dots,\boldsymbol{x}_N\}$ is the same point-set in $\overline{\Omega}$ on which the noisy scattered data of $f \in W_2^s(\Omega)$ is given, but in general this need not be so.

In most RBF approximation problems the so-called *native space* of the RBF plays an important role. For constructing the native space, initially the linear space

$$\mathcal{F}_\phi(\mathbb{S}^d) := \left\{\sum_{j=1}^{N} \alpha_j\, \phi(\cdot,\boldsymbol{x}_j) \;:\; \alpha_j \in \mathbb{R},\ \boldsymbol{x}_j \in \mathbb{S}^d,\ j = 1,2,\dots,N, \text{ and } N \in \mathbb{N}\right\}$$

of all possible linear combinations of shifted copies of $\phi(\cdot,\boldsymbol{x})$ is considered. This space $\mathcal{F}_\phi(\mathbb{S}^d)$ is a Pre-Hilbert space equipped with the inner product

$$\left(\sum_{j=1}^{N} \alpha_j\, \phi(\cdot,\boldsymbol{x}_j), \sum_{k=1}^{M} \beta_k\, \phi(\cdot,\boldsymbol{y}_k)\right)_\phi := \sum_{j=1}^{N}\sum_{k=1}^{M} \alpha_j\, \beta_k\, \phi(\boldsymbol{x}_j,\boldsymbol{y}_k). \tag{3.2}$$

The *native space* $\mathcal{N}_\phi(\mathbb{S}^d)$ is defined as the closure of $\mathcal{F}_\phi(\mathbb{S}^d)$ with respect to the norm $\|g\|_\phi := \sqrt{(g,g)_\phi}$, $g \in \mathcal{F}_\phi(\mathbb{S}^d)$, induced by the inner product (3.2). We denote the inner product and induced norm of the native space $\mathcal{N}_\phi(\mathbb{S}^d)$ also by $(\cdot,\cdot)_\phi$ and $\|\cdot\|_\phi$, respectively.

By construction, the native space $\mathcal{N}_\phi(\mathbb{S}^d)$ is a *reproducing kernel Hilbert space* with the reproducing kernel ϕ, that is, (i) $\phi(\boldsymbol{x},\boldsymbol{y}) = \phi(\boldsymbol{y},\boldsymbol{x})$ for all $\boldsymbol{x},\boldsymbol{y} \in \mathbb{S}^d$, (ii) $\phi(\cdot,\boldsymbol{x}) \in \mathcal{N}_\phi(\mathbb{S}^d)$ for all $\boldsymbol{x} \in \mathbb{S}^d$, and (iii) $(g,\phi(\cdot,\boldsymbol{x}))_\phi = g(\boldsymbol{x})$ for all $g \in \mathcal{N}_\phi(\mathbb{S}^d)$ and all $\boldsymbol{x} \in \mathbb{S}^d$.

For our particular approximation problem, we will choose the RBF such that its native space is the (global) Sobolev space $W_2^s(\mathbb{S}^d) = H^s(\mathbb{S}^d)$ with $s > d/2$. The native space norm $\|\cdot\|_\phi$ is then equivalent to the usual Sobolev space norm $\|\cdot\|_{W_2^s(\mathbb{S}^d)}$.

For more information on RBFs, consult [18], and for RBFs on the sphere, see [12]. For an introduction to reproducing kernel Hilbert spaces, see [3].

3.3 Local and global L_2 error estimates for RBF-based penalized least-squares approximation from noisy scattered data

Let Ω be either the sphere $\mathbb{S}^d$ or a nonempty connected open subset of $\mathbb{S}^d$ with Lipschitz-continuous boundary, contained in a hemisphere. Given noisy scattered data $f(\boldsymbol{x}_i) + \epsilon_i$, $i = 1, 2, \ldots, N$, of a function $f \in W_2^s(\Omega)$, where $s > d/2$, on a point-set $X_N = \{\boldsymbol{x}_1, \boldsymbol{x}_2, \ldots, \boldsymbol{x}_N\} \subseteq \overline{\Omega}$ of N distinct points, we choose an RBF ϕ such that its native space $\mathcal{N}_\phi(\mathbb{S}^d)$ (with norm $\|\cdot\|_\phi$) is $W_2^s(\mathbb{S}^d)$. For $\lambda > 0$, the *(RBF-based) penalized least-squares approximation* f_λ^ϵ of f is then defined as the *unique minimizer in* $W_2^s(\mathbb{S}^d)$ of the quadratic functional

$$\mu_\lambda(g) = \sum_{i=1}^{N} \left[g(\boldsymbol{x}_i) - (f(\boldsymbol{x}_i) + \epsilon_i)\right]^2 + \lambda \|g\|_\phi^2, \quad g \in W_2^s(\mathbb{S}^d). \tag{3.3}$$

The existence of a unique minimizer follows from the convexity of the minimized functional. For the case $\lambda = 0$, any interpolant in $W_2^s(\mathbb{S}^d)$ of the noisy scattered data is a minimizer, but there is a unique interpolant with minimal norm $\|\cdot\|_\phi$, which we will denote by f_0^ϵ.

We note that, even if we approximate a local function $f \in W_2^s(\Omega)$ defined only on the closure $\overline{\Omega}$ of the nonempty connected open subset Ω of $\mathbb{S}^d$ with Lipschitz-continuous boundary, we always seek an approximation f_λ^ϵ that is defined everywhere on $\mathbb{S}^d$. It can be shown that the restriction of f_λ^ϵ to $\overline{\Omega}$ equals the minimizer of a corresponding local minimization problem.

Surprisingly, for $\lambda > 0$, the unique minimizer in $W_2^s(\mathbb{S}^d)$ of (3.3) (and also the minimal $\|\cdot\|_\phi$ norm interpolant for $\lambda = 0$) is an RBF approximation of the form

$$f_\lambda^\epsilon = \sum_{j=1}^{N} \alpha_j \, \phi(\cdot, \boldsymbol{x}_j),$$

whose coefficient vector $\boldsymbol{\alpha} = (\alpha_1, \ldots, \alpha_N)^T \in \mathbb{R}^N$ satisfies

$$f_\lambda^\epsilon(\boldsymbol{x}_i) + \lambda\, \alpha_i = \sum_{j=1}^{N} \alpha_j \, \phi(\boldsymbol{x}_i, \boldsymbol{x}_j) + \lambda\, \alpha_i = f(\boldsymbol{x}_i) + \epsilon_i, \quad i = 1, \ldots, N.$$

Thus, for computing the approximation, we need to solve the linear system

$$(\boldsymbol{A} + \lambda \boldsymbol{I})\,\boldsymbol{\alpha} = \boldsymbol{f} + \boldsymbol{\epsilon}, \tag{3.4}$$

where $\boldsymbol{f} = (f(\boldsymbol{x}_1), f(\boldsymbol{x}_2), \ldots, f(\boldsymbol{x}_N))^T$, $\boldsymbol{\epsilon} = (\epsilon_1, \epsilon_2, \ldots, \epsilon_N)^T \in \mathbb{R}^N$, $\mathbf{A} = [\phi(\boldsymbol{x}_i, \boldsymbol{x}_j)]_{i,j=1,\ldots,N} \in \mathbb{R}^{N\times N}$, and where $\boldsymbol{I} \in \mathbb{R}^{N\times N}$ is the $N \times N$ identity matrix. Since ϕ is strictly positive definite, the matrix $\boldsymbol{A}$ is positive definite, and the linear system (3.4) therefore has for all $\lambda \geq 0$ a unique solution. As λ increases, the condition of the matrix $\boldsymbol{A}+\lambda\boldsymbol{I}$ improves.

Penalized least-squares approximation in the context of noisy data was already discussed by Wahba in [16]. For more background information on penalized least-squares approximation, we refer also to [1, 17].

Let us now consider the role of the *regularization parameter* (or *smoothing parameter*) $\lambda > 0$. For $\lambda > 0$ close to zero, much importance is given to fitting the data, and we are still close to the interpolation scenario. As $\lambda > 0$ increases, fitting the data becomes less and less important, and more and more weight is given to keeping the norm $\|f_\lambda^{\boldsymbol{\epsilon}}\|_\phi$ small, that is, to obtaining a smooth approximation. It is clear that $\lambda \geq 0$ needs to be chosen depending on the noise level.

Restricted to the N-dimensional approximation space

$$V_{X_N} := \left\{ \sum_{j=1}^{N} \alpha_j\, \phi(\cdot, \boldsymbol{x}_j) \;:\; \alpha_1, \alpha_2, \ldots, \alpha_N \in \mathbb{R} \right\},$$

the minimized functional (3.3) can be rewritten as

$$\widetilde{\mu}_\lambda(\boldsymbol{\alpha}) := \mu_\lambda\left(\sum_{j=1}^{N} \alpha_j\, \phi(\cdot, \boldsymbol{x}_j) \right) = \|\boldsymbol{A}\,\boldsymbol{\alpha} - (\boldsymbol{f} + \boldsymbol{\epsilon})\|_2^2 + \lambda\, \|\boldsymbol{\alpha}\|_{\boldsymbol{A}}^2 \tag{3.5}$$

with $\|\boldsymbol{\alpha}\|_{\boldsymbol{A}} := \sqrt{\boldsymbol{\alpha}^T \boldsymbol{A}\,\boldsymbol{\alpha}}$. (In (3.5), $\|\cdot\|_2$ denotes the Euclidean norm on $\mathbb{R}^N$.) From (3.5) it is obvious that the minimization problem (restricted to the finite dimensional approximation space V_{X_N}) can be interpreted as an instance of *Tikhonov regularization* (see [13, Section 2.2]). Hence, parameter choice strategies for Tikhonov regularization can be applied for choosing the regularization parameter λ, depending on the noise level.

In [10, 11], one *a posteriori* parameter choice strategy and some *a priori* parameter choice strategies were considered, but here we will only report on the *a posteriori* parameter choice strategy, namely *Morozov's discrepancy principle*: In the context given above and assuming furthermore that $\boldsymbol{f} + \boldsymbol{\epsilon} \neq \boldsymbol{0}$, it can be shown (see [10, Lemma 3.1]) that the *discrepancy functional*

$$J : \mathbb{R}_0^+ \to \mathbb{R}_0^+, \quad J(\lambda) := \sum_{i=1}^{N} [f_\lambda^{\boldsymbol{\epsilon}}(\boldsymbol{x}_i) - (f(\boldsymbol{x}_i) + \epsilon_i)]^2,$$

is continuous and strictly monotonic increasing with the range $\mathcal{R}(J) = [0, \|\boldsymbol{f} + \boldsymbol{\epsilon}\|_2^2[$, where $\|\boldsymbol{f} + \boldsymbol{\epsilon}\|_2^2 = \sum_{i=1}^{N} (f(\boldsymbol{x}_i) + \epsilon_i)^2$. Thus, if the noisy data satisfies $0 < \|\boldsymbol{\epsilon}\|_2 < \|\boldsymbol{f} + \boldsymbol{\epsilon}\|_2$, there exists a *unique regularization parameter* $\lambda > 0$ such that

$$J(\lambda) = \sum_{i=1}^{N} [f_\lambda^{\epsilon}(\boldsymbol{x}_i) - (f(\boldsymbol{x}_i) + \epsilon_i)]^2 = \sum_{i=1}^{N} \epsilon_i^2 = \|\boldsymbol{\epsilon}\|_2^2. \tag{3.6}$$

Morozov's discrepancy principle suggests to choose λ such that (3.6) is satisfied, since the squared norm of the ℓ_2 discrepancy of the approximation f_λ^{ϵ} from the noisy scattered data then exactly corresponds to the noise level $\|\boldsymbol{\epsilon}\|_2^2$ (which would also be the case if we could recover f itself).

In [10, Theorem 4.1] and [11, Theorem 6], the following $L_2(\Omega)$ error estimates for the approximation f_λ^{ϵ}, with λ chosen with Morozov's discrepancy principle, have been proved.

Theorem 3.1. *Let $d \geq 2$ and $s > d/2$, and let Ω be either $\mathbb{S}^d$ or a nonempty connected open subset of $\mathbb{S}^d$ with Lipschitz-continuous boundary, contained in an open hemisphere. Let ϕ be an RBF whose native space $\mathcal{N}_\phi(\mathbb{S}^d)$ with norm $\|\cdot\|_\phi$ is the Sobolev space $W_2^s(\mathbb{S}^d)$. For a point-set $X_N = \{\boldsymbol{x}_1, \boldsymbol{x}_2, \ldots, \boldsymbol{x}_N\} \subseteq \overline{\Omega}$ of N distinct points in $\overline{\Omega}$ and a function $f \in W_2^s(\Omega)$ with values $\boldsymbol{f} = (f(\boldsymbol{x}_1), f(\boldsymbol{x}_2), \ldots, f(\boldsymbol{x}_N))^T$ on X_N, and given a vector of deterministic noise $\boldsymbol{\epsilon} = (\epsilon_1, \epsilon_2, \ldots, \epsilon_N)^T \neq \boldsymbol{0}$ and given $\lambda > 0$, let f_λ^{ϵ} denote the uniquely determined minimizer in $W_2^s(\mathbb{S}^d)$ of the quadratic functional*

$$\mu_\lambda(g) := \sum_{i=1}^{N} [g(\boldsymbol{x}_i) - (f(\boldsymbol{x}_i) + \epsilon_i)]^2 + \lambda \|g\|_\phi^2, \quad g \in W_2^s(\mathbb{S}^d).$$

Then there exist positive constants $c_{d,s}$, $c'_{d,s}$ and $\hat{c}_{d,s}$ (dependent on d, s, Ω and ϕ) such that, if the point-set X_N satisfies $h_{X_N,\overline{\Omega}} \leq \hat{c}_{s,d}$, then the following result holds true for all $f \in W_2^s(\Omega)$ and all $\boldsymbol{\epsilon} \in \mathbb{R}^N \setminus \{\boldsymbol{0}\}$: If $\|\boldsymbol{\epsilon}\|_2 < \|\boldsymbol{f} + \boldsymbol{\epsilon}\|_2$, there exists a unique $\lambda > 0$ such that

$$\sum_{i=1}^{N} [f_\lambda^{\epsilon}(\boldsymbol{x}_i) - (f(\boldsymbol{x}_i) + \epsilon_i)]^2 = \sum_{i=1}^{N} \epsilon_i^2 = \|\boldsymbol{\epsilon}\|_2^2. \tag{3.7}$$

For this choice of λ, the approximation f_λ^{ϵ} satisfies the following $L_2(\Omega)$ error estimate:

$$\|f_\lambda^{\epsilon} - f\|_{L_2(\Omega)} \leq c_{d,s}\, h_{X_N,\overline{\Omega}}^s \|f\|_{W_2^s(\Omega)} + c'_{d,s}\, h_{X_N,\overline{\Omega}}^{d/2} \|\boldsymbol{\epsilon}\|_2. \tag{3.8}$$

The error estimate (3.8) is *order optimal* in the following sense:

- The term $h_{X_N,\overline{\Omega}}^s \|f\|_{W_2^s(\Omega)}$ has the correct order of the mesh norm $h_{X_N,\overline{\Omega}}$ for an estimate of the $L_2(\Omega)$ norm against the $W_2^s(\Omega)$ norm.
- For a sequence $(\widetilde{X}_{N_m})_{m\in\mathbb{N}}$ (with $\lim_{m\to\infty} N_m = \infty$) of "well-distributed" point-sets $\widetilde{X}_{N_m} = \{\boldsymbol{x}_1^{(m)}, \boldsymbol{x}_2^{(m)}, \ldots, \boldsymbol{x}_{N_m}^{(m)}\} \subseteq \overline{\Omega}$ of N_m distinct points, where "well-distributed" means that $h_{\widetilde{X}_{N_m},\overline{\Omega}} \asymp N_m^{-1/d}$, we obtain

$$h_{\widetilde{X}_{N_m},\overline{\Omega}}^{d/2} \|\boldsymbol{\epsilon}\|_2 \asymp \frac{1}{\sqrt{N_m}} \|\boldsymbol{\epsilon}\|_2.$$

(For two positive sequences, $(a_m)_{m\in\mathbb{N}}$ and $(b_m)_{m\in\mathbb{N}}$, the notation $a_m \asymp b_m$ means that there exist positive constants c_1, c_2 such that $c_1\, a_m \le b_m \le c_2\, a_m$ for all $m \in \mathbb{N}$.) As the $L_2(\Omega)$ error of the approximation can never be smaller than the root-mean-square error $\frac{1}{\sqrt{N_m}}\, \|\boldsymbol{\epsilon}\|_2$ of the data, the second term is also order optimal for "well-distributed" point-sets.

3.4 Main ideas in the proof of the L_2 error estimates

Proof. The starting point for the proof is an L_2 sampling inequality (see [10, Theorem 5.1] for $\Omega = \mathbb{S}^d$ and [11, Theorem 2] for Ω a nonempty connected open subset of $\mathbb{S}^d$ with Lipschitz-continuous boundary, contained in an open hemisphere) which we apply to $f_\lambda^{\boldsymbol{\epsilon}} - f$,

$$\|f_\lambda^{\boldsymbol{\epsilon}} - f\|_{L_2(\Omega)} \le \widetilde{C}_{d,s}\left[h^s_{X_N,\overline{\Omega}}\, \|f_\lambda^{\boldsymbol{\epsilon}} - f\|_{W_2^s(\Omega)} + h^{d/2}_{X_N,\overline{\Omega}} \left(\sum_{i=1}^N [f_\lambda^{\boldsymbol{\epsilon}}(\mathbf{x}_i) - f(\mathbf{x}_i)]^2 \right)^{1/2} \right]. \tag{3.9}$$

This sampling inequality holds for all functions in $W_2^s(\Omega)$ under the assumption that $h_{X_N,\overline{\Omega}} \le \widehat{C}_{d,s}$. The L_2 sampling inequalities [10, Theorem 5.1] and [11, Theorem 2] were proved by lifting a special case of the general sampling inequality [2, Theorem 4.1] with a suitably chosen atlas for $\mathbb{S}^d$ from a suitable subset of $\mathbb{R}^d$ with Lipschitz-continuous boundary onto the sphere $\mathbb{S}^d$.

In the second term in (3.9), we add and subtract ϵ_i, and then use $(a+b)^2 \le 2\,(a^2+b^2)$ for all $a, b \in \mathbb{R}$ and finally use the fact that λ is chosen with Morozov's discrepancy principle (3.7):

$$\begin{aligned}\sum_{i=1}^N [f_\lambda^{\boldsymbol{\epsilon}}(\mathbf{x}_i) - f(\mathbf{x}_i)]^2 &= \sum_{i=1}^N [f_\lambda^{\boldsymbol{\epsilon}}(\mathbf{x}_i) - (f(\mathbf{x}_i) + \epsilon_i) + \epsilon_i]^2 \\ &\le 2\left(\sum_{i=1}^N [f_\lambda^{\boldsymbol{\epsilon}}(\mathbf{x}_i) - (f(\mathbf{x}_i) + \epsilon_i)]^2 + \sum_{i=1}^N \epsilon_i^2 \right) \\ &= 2\left(\sum_{i=1}^N \epsilon_i^2 + \sum_{i=1}^N \epsilon_i^2 \right) \\ &= 4\|\boldsymbol{\epsilon}\|_2^2.\end{aligned} \tag{3.10}$$

To estimate $\|f_\lambda^{\boldsymbol{\epsilon}} - f\|_{W_2^s(\Omega)}$, we note that unlike $f \in W_2^s(\Omega)$, the approximation $f_\lambda^{\boldsymbol{\epsilon}} \in W_2^s(\mathbb{S}^d)$ is always defined on the whole sphere $\mathbb{S}^d$ and satisfies $\|f_\lambda^{\boldsymbol{\epsilon}}\|_{W_2^s(\Omega)} \le \|f_\lambda^{\boldsymbol{\epsilon}}\|_{W_2^s(\mathbb{S}^d)}$. In the first term in (3.9), we use the triangle inequality, and after that $\|f_\lambda^{\boldsymbol{\epsilon}}\|_{W_2^s(\Omega)} \le \|f_\lambda^{\boldsymbol{\epsilon}}\|_{W_2^s(\mathbb{S}^d)} \le c'\, \|f_\lambda^{\boldsymbol{\epsilon}}\|_\phi$ (with the second estimate valid, because $\|\cdot\|_{W_2^s(\mathbb{S}^d)}$ and $\|\cdot\|_\phi$ are equivalent norms), giving

$$\begin{aligned}\|f_\lambda^{\boldsymbol{\epsilon}} - f\|_{W_2^s(\Omega)} &\le \|f_\lambda^{\boldsymbol{\epsilon}}\|_{W_2^s(\Omega)} + \|f\|_{W_2^s(\Omega)} \\ &\le \|f_\lambda^{\boldsymbol{\epsilon}}\|_{W_2^s(\mathbb{S}^d)} + \|f\|_{W_2^s(\Omega)} \\ &\le c' \, \|f_\lambda^{\boldsymbol{\epsilon}}\|_\phi + \|f\|_{W_2^s(\Omega)}. \end{aligned} \tag{3.11}$$

To estimate $\|f_\lambda^{\boldsymbol{\epsilon}}\|_\phi$, we will exploit the optimality of $f_\lambda^{\boldsymbol{\epsilon}}$, that is, $\mu_\lambda(f_\lambda^{\boldsymbol{\epsilon}}) \le \mu_\lambda(g)$ for all $g \in W_2^s(\mathbb{S}^d)$. We start with

$$\begin{aligned}\|f_\lambda^{\boldsymbol{\epsilon}}\|_\phi^2 = \frac{1}{\lambda}(\lambda \, \|f_\lambda^{\boldsymbol{\epsilon}}\|_\phi^2) &= \frac{1}{\lambda}\left(\mu_\lambda(f_\lambda^{\boldsymbol{\epsilon}}) - \sum_{i=1}^N [f_\lambda^{\boldsymbol{\epsilon}}(\mathbf{x}_i) - (f(\mathbf{x}_i) + \epsilon_i)]^2 \right) \\ &= \frac{1}{\lambda}\left(\mu_\lambda(f_\lambda^{\boldsymbol{\epsilon}}) - \sum_{i=1}^N \epsilon_i^2 \right) \\ &= \frac{1}{\lambda}(\mu_\lambda(f_\lambda^{\boldsymbol{\epsilon}}) - \|\boldsymbol{\epsilon}\|_2^2), \end{aligned} \tag{3.12}$$

where we have used (3.7) again in the second line.

Ideally, we would like to replace $f_\lambda^{\boldsymbol{\epsilon}}$ in (3.12) by f, giving $\mu_\lambda(f_\lambda^{\boldsymbol{\epsilon}}) \le \mu_\lambda(f)$. However, this is only possible for $\Omega = \mathbb{S}^d$, since otherwise f is only in $W_2^s(\Omega)$, but not in $W_2^s(\mathbb{S}^d)$. To overcome this problem, we use the Sobolev extension $E_s(f) \in W_2^s(\mathbb{S}^d)$ of $f \in W_2^s(\Omega)$, given by [11, Theorem 3]. [11, Theorem 3] is a suitably constructed Sobolev extension theorem, guaranteeing that the extension $E_s(f)$ satisfies $E_s(f) \equiv f$ on $\overline{\Omega}$ and $\|E_s(f)\|_{W_2^s(\mathbb{S}^d)} \le c \, \|f\|_{W_2^s(\Omega)}$ with a positive constant c independent of $f \in W_2^s(\Omega)$. Using the equivalence of the norms $\|\cdot\|_{W_2^s(\mathbb{S}^d)}$ and $\|\cdot\|_\phi$, we then have $\|E_s(f)\|_\phi \le c'' \, \|E_s(f)\|_{W_2^s(\mathbb{S}^d)} \le c'' \, c \, \|f\|_{W_2^s(\Omega)}$.

In the following, let $\widetilde{E}_s(f) := f$ for $\Omega = \mathbb{S}^d$; and for Ω a nonempty connected open subset of $\mathbb{S}^d$ with Lipschitz-continuous boundary, contained in an open hemisphere, let $\widetilde{E}_s(f) := E_s(f)$ be the extension $E_s(f) \in W_2^s(\mathbb{S}^d)$ of f given by the Sobolev extension theorem [11, Theorem 3]. Then $\widetilde{E}_s(f) \equiv f$ on $\overline{\Omega}$ and $\|\widetilde{E}_s(f)\|_\phi \le \widetilde{c} \, \|f\|_{W_2^s(\Omega)}$ with $\widetilde{c} := c'' \max\{c, 1\}$. Using in (3.12) $\mu_\lambda(f_\lambda^{\boldsymbol{\epsilon}}) \le \mu_\lambda(\widetilde{E}_s(f))$ and subsequently $(\widetilde{E}_s(f))(\mathbf{x}_i) = f(\mathbf{x}_i)$ for $i = 1, 2, \ldots, N$ yields

$$\begin{aligned}\|f_\lambda^{\boldsymbol{\epsilon}}\|_\phi^2 &= \frac{1}{\lambda}(\mu_\lambda(f_\lambda^{\boldsymbol{\epsilon}}) - \|\boldsymbol{\epsilon}\|_2^2) \\ &\le \frac{1}{\lambda}(\mu_\lambda(\widetilde{E}_s(f)) - \|\boldsymbol{\epsilon}\|_2^2) \\ &= \frac{1}{\lambda}\left(\sum_{i=1}^N [(\widetilde{E}_s(f))(\mathbf{x}_i) - (f(\mathbf{x}_i) + \epsilon_i)]^2 + \lambda \, \|\widetilde{E}_s(f)\|_\phi^2 - \|\boldsymbol{\epsilon}\|_2^2 \right) \\ &\le \frac{1}{\lambda}\left(\sum_{i=1}^N \epsilon_i^2 + \lambda \, \widetilde{c}^2 \, \|f\|_{W_2^s(\Omega)}^2 - \|\boldsymbol{\epsilon}\|_2^2 \right) \\ &= \widetilde{c}^2 \, \|f\|_{W_2^s(\Omega)}^2. \end{aligned} \tag{3.13}$$

Applying (3.13) in (3.11) yields

$$\begin{aligned}\|f_\lambda^{\epsilon} - f\|_{W_2^s(\Omega)} &\le c' \, \|f_\lambda^{\epsilon}\|_{\phi} + \|f\|_{W_2^s(\Omega)} \\ &\le c' \, \widetilde{c} \, \|f\|_{W_2^s(\Omega)} + \|f\|_{W_2^s(\Omega)} \\ &\le (c' \, \widetilde{c} + 1) \, \|f\|_{W_2^s(\Omega)}. \end{aligned} \tag{3.14}$$

From (3.9), (3.10), and (3.14)

$$\|f_\lambda^{\epsilon} - f\|_{L_2(\Omega)} \le \widetilde{c}_{d,s} \big[(c' \, \widetilde{c} + 1) \, h_{X_N,\overline{\Omega}}^s \, \|f\|_{W_2^s(\Omega)} + 2 \, h_{X_N,\overline{\Omega}}^{d/2} \, \|\epsilon\|_2 \big],$$

which proves the $L_2(\Omega)$ error estimate (3.8). □

It should be noted here that a fair amount of rather technical arguments in [10, 11] goes into lifting the L_2 sampling inequality from open sets with Lipschitz-continuous boundary in $\mathbb{R}^d$ onto $\mathbb{S}^d$, as well as into the construction of the particular Sobolev extension theorem in [11], tailored to our needs. Moreover, in [11] the choice of the atlas and the associated partition of unity for $\mathbb{S}^d$ as a d-dimensional differentiable submanifold of $\mathbb{R}^{d+1}$, as well as the Sobolev extension theorem, providing an extension with local support tailored to the atlas and partition of unity, are crucial components to extend the results from the global to the local case. For background information on Sobolev spaces and Sobolev extension theorems, the reader is referred to [15].

In both papers [10, 11], the fact that Sobolev spaces $W_2^s(\mathbb{S}^d)$ on the sphere can be defined either with an atlas and an associated partition of unity for $\mathbb{S}^d$ as a d-dimensional differentiable submanifold of $\mathbb{R}^{d+1}$, or intrinsically to the sphere via the decay of the coefficients in Fourier series expansions (in terms of spherical harmonics), is essential. These two different definitions of $W_2^s(\mathbb{S}^d)$ yield different but equivalent norms (see [14, Section 7.3]). Whereas the definition of $W_2^s(\mathbb{S}^d)$ intrinsically to the sphere is not explicitly visible in this review paper, in [10, 11] it occurs in the context of the native space of the RBF ϕ, once the series expansion of ϕ, in terms of certain Jacobi polynomials, is introduced.

Bibliography

[1] P. Alfeld, M. Neamtu, and L. L. Schumaker. Fitting scattered data on sphere-like surfaces using spherical splines. J. Comput. Appl. Math., 73 (1996), 5–43.

[2] R. Arcangéli, M. C. López de Silanes, and J. J. Torrens. An extension of a bound for functions in Sobolev spaces, with applications to (m, s)-spline interpolation and smoothing. Numer. Math., 107 (2007), 181–211.

[3] N. Aronszajn. Theory of reproducing kernels. Trans. Amer. Math. Soc., 68 (3) (1950), 337–404.

[4] W. Freeden. Multiscale Modelling of Spaceborne Geodata. B. G. Teubner, Stuttgart, Leipzig, 1999.

[5] W. Freeden, T. Gervens, and M. Schreiner. Constructive Approximation on the Sphere (with Applications to Geomathematics). Clarendon Press, Oxford, 1998.

[6] W. Freeden and V. Michel. Multiscale Potential Theory (with Applications to Geoscience). Birkhäuser, Boston, 2004.

[7] W. Freeden and M. Z. Nashed (eds.). Handbook of Mathematical Geodesy (Functional Analytic and Potential Theoretic Methods). Birkhäuser, Springer International Publishing AG, part of Springer Nature, 2018.

[8] W. Freeden, M. Z. Nashed, and M. Schreiner. Spherical Sampling. Birkhäuser, Springer International Publishing AG, part of Springer Nature, 2018.

[9] W. Freeden, M. Z. Nashed, and T. Sonar (eds.). Handbook of Geomathematics. Springer-Verlag, Berlin, Heidelberg, 2010.

[10] K. Hesse, I. H. Sloan, and R. S. Womersley. Radial basis function approximation of noisy scattered data on the sphere. Numer. Math., 137 (2017), 579–605.

[11] K. Hesse, I. H. Sloan, and R. S. Womersley. Local RBF-based penalized least-squares approximation on the sphere with noisy scattered data. Submitted, 2019.

[12] S. Hubbert, Q. T. Le Gia, and T. M. Morton. Spherical Radial Basis Functions, Theory and Applications. Springer Cham, Heidelberg, New York, Dordrecht, London, 2015.

[13] A. Kirsch. An Introduction to the Mathematical Theory of Inverse Problems. Springer-Verlag, New York, 1996.

[14] J. L. Lions and E. Magenes. Non-Homogeneous Boundary Value Problems and Applications. Vol. 1. Springer-Verlag, Berlin, Heidelberg, New York, 1972.

[15] W. McLean. Strongly Elliptic Systems and Boundary Integral Equations. Cambridge University Press, Cambridge, 2000.

[16] G. Wahba. Spline Models for Observational Data. CBMS-NSF Regional Conference Series in Applied Mathematics, Vol. 59, SIAM, Philadelphia, PA, 1990.

[17] M. von Golitschek and L. L. Schumaker. Data fitting by penalized least squares. In: Algorithms for Approximation, II (Shrivenham, 1988), 210–227, Chapman and Hall, London, 1990.

[18] H. Wendland. Scattered Data Approximation. Cambridge University Press, Cambridge, 2005.

Aicke Hinrichs, David Krieg, Erich Novak, Joscha Prochno, and Mario Ullrich

4 On the power of random information

Abstract: We study approximation and integration problems and compare the quality of optimal information with the quality of random information. For some problems random information is almost optimal and for some other problems random information is much worse than optimal information. We prove new results and give a short survey of known results.

Keywords: Random information, L_p approximation, numerical integration, random intersections

MSC 2010: 65Y20, 52A23, 65D15, 65D32, 68Q87

4.1 Introduction

Optimal algorithms use optimal information about the problem instances. Quite often, we do not have access to optimal information. One reason may be that we do not know how to choose the most significant measurements. In this paper, we assume that information comes in randomly, which is a standard assumption in learning theory and uncertainty quantification. We ask the following question:

What is the typical quality of random information?

Of course, random information cannot be better than optimal information. In many cases, however, it will turn out that random information is practically as good as op-

Acknowledgement: The authors would like to thank the Isaac Newton Institute for Mathematical Sciences for support and hospitality during the programme "Approximation, sampling and compression in data science" when work on this paper was undertaken. This work was supported by EPSRC Grant Number EP/R014604/1. A. Hinrichs, D. Krieg and J. Prochno are supported by the Austrian Science Fund (FWF) Projects F5513-N26 and F5508-N26, which are part of the Special Research Program "Quasi-Monte Carlo Methods: Theory and Applications". A. Hinrichs was also supported by a grant from the Simons Foundation. J. Prochno is also supported by a Visiting Professor Fellowship of the Ruhr University Bochum and its research school, PLUS.

Aicke Hinrichs, David Krieg, Mario Ullrich, Institut für Analysis, Johannes Kepler Universität Linz, Altenbergerstrasse 69, 4040 Linz, Austria, e-mails: aicke.hinrichs@jku.at, david.krieg@jku.at, mario.ullrich@jku.at
Erich Novak, Mathematisches Institut, Friedrich-Schiller-Universität Jena, Ernst-Abbe-Platz 2, 07743 Jena, Germany, e-mail: erich.novak@uni-jena.de
Joscha Prochno, Institut für Mathematik & Wissenschaftliches Rechnen, Karl-Franzens-Universität, Heinrichstraße 36, 8010 Graz, Austria, e-mail: joscha.prochno@uni-graz.at

https://doi.org/10.1515/9783110635461-004

timal information, whereas in other situations random information is a little worse. Sometimes, random information is completely useless.

To phrase our question precisely, we need to clarify how we measure the quality of information and what we mean by random. The first is done with the so-called radius of information, which is the worst-case error of the best algorithm that uses nothing but the given information, and the a priori knowledge about the problem instance. Random information, on the other hand, shall be obtained from a certain number of independent measurements that all follow the same law. We study the question for two different natural choices of uniform distributions.

Let us go a little more into detail. A linear problem is given by a linear solution operator S that maps from a convex and symmetric subset F of a normed space to a normed space G, and a class Λ of continuous linear functionals on F, the class of admissible measurements. We may think of an integration problem, where $S(f)$ is the integral of a function f, or a recovery problem, where S is an embedding. One wants to approximate the solution $S(f)$ for unknown $f \in F$ based on n of these measurements such that we can guarantee a small error with respect to the norm in G. We refer the reader to [30, 37] for a detailed exposition. An information mapping has the form

$$N_n : F \to \mathbb{R}^n, \quad N_n(f) = (L_1(f), \dots, L_n(f)), \tag{4.1}$$

where $L_1, \dots, L_n \in \Lambda$. The power or quality of the information mapping is measured by the radius of information. This is the worst-case error of the best algorithm $A_n = \phi \circ N_n$ based on this information, i. e.,

$$r(N_n) = \inf_{\phi:\mathbb{R}^n \to G} \sup_{f \in F} \|S(f) - \phi(N_n(f))\|_G. \tag{4.2}$$

We say that information is optimal, and write N_n^* instead of N_n, if

$$r(N_n^*) = \inf_{N_n} r(N_n). \tag{4.3}$$

Here, we study random information of the form (4.1), where the random functionals $L_1, \dots, L_n \in \Lambda$ are independent and identically distributed. The goal is to compare

$$r(N_n^*) \quad \text{vs.} \quad \mathbb{E}[r(N_n)],$$

the radius of optimal information and the expected radius of random information.

If the infimum and the expected value are comparable, this means that there are many good algorithms based on many different information mappings. In this case, optimal information and therefore optimal algorithms are not very special. On the other hand, if the infimum is significantly smaller than the expected value, this means that optimal information is indeed special. It seems to be an interesting characteristic of a problem whether optimal information is special or not. Of course, the results of our comparisons heavily depend on the distribution underlying the measurements. Whereas the question may be interesting for many distributions, we feel that there often is a natural choice. If the distribution only depends on the class Λ of admissible measurements, collecting random information might even be a good idea if optimal information is available. It may happen that we do not loose much in terms of the radius, but gain the following useful properties:

- Since the distribution is independent of n, it is easy to increase the number of measurements if our current approximation is not yet satisfactory.
- The information can be used for many different input classes F, solution operators S, and target spaces G. In that sense it is *universal*.

The second property does not mean that the corresponding algorithm $A_n = \phi \circ N_n$ is universal. The optimal choice of ϕ may depend on F, S, and G. It is well known, see again [37], that any interpolatory algorithm is optimal up to a factor two.

We note that examples of the sort "*random information is good*" can be deduced from work using the probabilistic method to prove upper bounds of $r(N_n^*)$. Here, one uses upper bounds of $\mathbb{E}[r(N_n)]$ and the trivial inequality $r(N_n^*) \leq \mathbb{E}[r(N_n)]$, see [17, 30, 32–34] for many such results. The idea is to introduce a random family of algorithms and to show that the expected worst-case error is small. This is used to obtain the existence of good algorithms. However, it actually implies that *most* of the algorithms in that family are good. Therefore, the expected radius of the random information underlying these algorithms must be small as well. We note that for this approach only upper bounds of $\mathbb{E}[r(N_n)]$ are needed, and lower bounds are usually not studied.

4.2 Uniformly distributed function evaluations

If the input class F consists of functions, one natural restriction of the class of information functionals is so called *standard information* consisting of function evaluations. Moreover, if the domain of definition of functions in F carries a probability measure, a natural source of random information are function values $f(x)$, with x distributed according to the probability measure. In this section, we analyze this setting for different input classes F of smooth functions defined on the unit cube $[0,1]^d$. Then the random information consists of function values $f(x)$, with x uniformly distributed in $[0,1]^d$.

4.2.1 Integration and approximation in univariate Sobolev spaces of smoothness 1

We consider the L_q-approximation problem

$$\mathrm{APP}\colon W_p^1([0,1]) \to L_q([0,1]),$$

given by $\mathrm{APP}(f) = f$ using function values as information. The Sobolev spaces $W_p^1([0,1])$ are usually normed by

$$\|f\|_{W_p^1([0,1])} := \left(\|f\|_p^p + \|f'\|_p^p\right)^{1/p}.$$

It is well known that the optimal information is given by

$$N_n^*(f) = (f(1/n), f(3/n), \dots, f((2n-1)/n)),$$

i. e., point evaluations at equidistant points. The minimal radius of information then satisfies

$$r(N_n^*) \asymp n^{-1+(1/p-1/q)_+},$$

see (4.6).

Here and later, we use asymptotic notation as follows: For sequences (a_n) and (b_n) of positive real numbers, we write $a_n \preccurlyeq b_n$, and $b_n \succcurlyeq a_n$ to indicate that there exists a constant $C \in (0,\infty)$ such that $a_n \le C\, b_n$ for all n. We shall write $a_n \asymp b_n$ if $a_n \preccurlyeq b_n$ and $b_n \preccurlyeq a_n$. Sometimes we use similar notation for double sequences.

We will show that information given by n independent and uniformly distributed points in $[0,1]$ is as good as optimal information provided that $p > q$, whereas in the case $p \le q$, there is a loss of a logarithmic factor.

Theorem 4.1. *Consider L_q-approximation of functions from $W_p^s([0,1])$ with $1 \le p, q \le \infty$, using function values at uniformly distributed points in $[0,1]$.*

$$\mathbb{E}[r(N_n)] \asymp \begin{cases} r(N_n^*) \asymp n^{-1} & \text{if } p > q \\ r(N_{n/\log n}^*) \asymp \left(\frac{n}{\log n}\right)^{-1+1/p-1/q} & \text{if } p \le q. \end{cases}$$

The proof of Theorem 4.1 is based on the following formula for the radius of information $r(N_n)$ for information $N_n(f) = (f(x_1), f(x_2), \dots, f(x_n))$, where we assume $0 = x_0 \le x_1 \le \cdots \le x_n \le x_{n+1} = 1$.

Lemma 4.2.

$$r(N_n) \asymp \begin{cases} \left(\sum_{i=0}^n (x_{i+1}-x_i)^{(pq+p-q)/(p-q)}\right)^{1/q-1/p} & \text{if } p > q \\ \max_{0\le i\le n}(x_{i+1}-x_i)^{1-1/p+1/q} & \text{if } p \le q. \end{cases}$$

Proof. Note that in this case the radius of information satisfies

$$r(N_n) \asymp \sup_{\substack{f\in W_p^1([0,1])\\ N_n(f)=0}} \frac{\|f\|_q}{\|f\|_{W_p^1}},$$

with an equivalence constant of at most 2, see [30, Section 4.2]. We start with the lower bound in the case $p \le q$. Let $i_0 \in \{0,1,\dots,n\}$ be such that

$$\ell = x_{i_0+1} - x_{i_0} = \max_{0\le i\le n}(x_{i+1} - x_i),$$

and consider a continuous piecewise linear function $f : [0,1] \to \mathbb{R}$ that is zero outside of the interval (x_{i_0}, x_{i_0+1}), satisfies $f'(x) = 1$ in the left half of this interval, and $f'(x) = -1$ in the right half. Then

$$N_n(f) = (0,\dots,0), \quad \|f\|_q \asymp \ell^{1+1/q}, \quad \text{and} \quad \|f\|_{W_p^1} \asymp \ell^{1/p}$$

show that $r(N_n) \succcurlyeq \ell^{1-1/p+1/q}$.

We turn to the lower bound in the case $p > q$. In this case, let $f : [0,1] \to \mathbb{R}$ be a continuous piecewise linear function satisfying $f(x_i) = 0$ for $i = 0,1,\dots,n+1$ and $f'(x) = \ell_i^\alpha$ in the left half of each interval $[x_i, x_{i+1}]$ and $f'(x) = -\ell_i^\alpha$ in the right half, where $\ell_i = x_{i+1} - x_i$ and $\alpha = q/(p-q)$. Then $N_n(f) = (0,\dots,0)$, and

$$\|f\|_q \asymp \left(\sum_{i=0}^n \ell_i^{(pq+p-q)/(p-q)}\right)^{1/q} \quad \text{and} \quad \|f\|_{W_p^1} \asymp \left(\sum_{i=0}^n \ell_i^{(pq+p-q)/(p-q)}\right)^{1/p}$$

show that $r(N_n) \succcurlyeq (\sum_{i=0}^n \ell_i^{(pq+p-q)/(p-q)})^{1/q-1/p}$.

Finally, we prove the upper bound. Observe that $\|f\|_\infty \le \|f'\|_1$ holds for any function $f \in W_1^1([0,1])$ with $f(0) = 0$. Consequently, $\|f\|_q \le \|f'\|_p$ holds for any function $f \in W_p^1([0,1])$ with $f(0) = 0$. By scaling, it follows that $\|f\|_q \le (b-a)^{1-1/p+1/q}\|f'\|_p$ holds for any function $f \in W_p^1([a,b])$ satisfying either $f(a) = 0$ or $f(b) = 0$. Let now $f \in W_p^1([0,1])$ be such that $N_n(f) = (f(x_1), f(x_2), \dots, f(x_n)) = (0,\dots,0)$ and $\|f'\|_p = 1$. Let f_i be the restriction of f to the interval $[x_i, x_{i+1}]$. Then

$$\|f\|_q = \left(\sum_{i=0}^n \|f_i\|_q^q\right)^{1/q} \le \left(\sum_{i=0}^n (x_{i+1} - x_i)^{q-q/p+1}\|f_i'\|_p^q\right)^{1/q}.$$

In the case $p \le q$, let again ℓ be the length of the largest of the intervals $[x_i, x_{i+1}]$. It follows that

$$\begin{aligned}\|f\|_q &\le \ell^{1-1/p+1/q}\left(\sum_{i=0}^n \|f_i'\|_p^q\right)^{1/q} \le \ell^{1-1/p+1/q}\left(\sum_{i=0}^n \|f_i'\|_p^p\right)^{1/p}\\ &= \ell^{1-1/p+1/q}\|f'\|_p = \ell^{1-1/p+1/q}.\end{aligned}$$

This implies $r(N_n) \leq e^{1-1/p+1/q}$ as claimed. In the case $p > q$, let again $\ell_i = x_{i+1} - x_i$ and $s > 0$ be given by $\frac{1}{s} = \frac{1}{q} - \frac{1}{p}$. Using Hölder's inequality, we arrive at

$$\|f\|_q \leq \left(\sum_{i=0}^{n} \ell_i^{s+1}\right)^{1/s} \left(\sum_{i=0}^{n} \|f_i'\|_p^p\right)^{1/p} = \left(\sum_{i=0}^{n} \ell_i^{s+1}\right)^{1/s} \|f'\|_p$$
$$= \left(\sum_{i=0}^{n} \ell_i^{s+1}\right)^{1/s}.$$

This implies the claimed upper bound also when $p > q$. □

Proof of Theorem 4.1. In the case $p \leq q$, the proof follows from Lemma 4.2 and the fact that, in expectation, the largest gap among n independent and uniformly distributed points in $[0,1]$ is of order $\log(n)/n$. This is a simple consequence of Lemma 4.7. In the case $p > q$, the proof follows from

$$\mathbb{E}\left[\left(\sum_{i=0}^{n} \ell_i^{s+1}\right)^{1/s}\right] \asymp \frac{1}{n}$$

for any $s > 0$, where the ℓ_i are again the spacings of independent and uniformly distributed points in $[0,1]$. In fact, the lower bound follows trivially, since the expected radius of random information is larger than the radius of optimal information. To prove the upper bound, observe first that the function $s \mapsto (\sum_{i=0}^{n} \ell_i^{s+1})^{1/s}$ is nondecreasing, since $\sum_{i=0}^{n} \ell_i = 1$, and that Jensen's inequality implies

$$\mathbb{E}\left[\left(\sum_{i=0}^{n} \ell_i^{s+1}\right)^{1/s}\right] \leq \left(\mathbb{E}\left[\sum_{i=0}^{n} \ell_i^{s+1}\right]\right)^{1/s} \quad \text{whenever } s \geq 1.$$

Hence, it is enough to show that

$$\mathbb{E}\left[\sum_{i=0}^{n} \ell_i^{s+1}\right] \asymp n^{-s}$$

for positive integers s. It was proved in [9] that

$$\mathbb{E}\left[\sum_{i=0}^{n} h(\ell_i)\right] = n(n+1)\int_0^1 (1-r)^{n-1} h(r)\,\mathrm{d}r$$

for any integrable function $h : [0,1] \to \mathbb{R}$. With $h(r) = r^{s+1}$, we obtain

$$\mathbb{E}\left[\sum_{i=0}^{n} \ell_i^{s+1}\right] = n(n+1)B(s+2,n) = \frac{(n+1)!(s+1)!}{(n+s+1)!} \asymp n^{-s},$$

where $B(x,y)$ is the Beta function. This completes the proof. □

Remark 4.3. For the classes studied, i. e., when the smoothness is one, the radius of information $r(N_n)$ for the integration problem with $S(f) = \int_0^1 f(x)\,dx$ is the same as for the L_1-approximation problem. Hence, for the integration problem for functions in $W_p^1([0,1])$, we have

$$\mathbb{E}[r(N_n)] \asymp \begin{cases} r(N_n^*) \asymp n^{-1} & \text{if } p > 1 \\ r(N_{n/\log n}^*) \asymp (\frac{n}{\log n})^{-1} & \text{if } p = 1. \end{cases}$$

Some cases of Theorem 4.1 are known, sometimes in a slightly different language, since authors had in mind discrepancy (with optimal weights) instead of the integration problem. The star discrepancy for $d = 1$, i. e., $p = q = 1$, was studied in [39]. Here we have $r(N_n^*) = \frac{1}{2n}$. From Corollary 3.3 and Theorem 3.5 of [39] we know that

$$\frac{c \log n}{n} \le \mathbb{E}[r(N_n)] \le \frac{8 \log n}{n}$$

for some $c \in (0, \infty)$. Therefore, random information is good, but not optimal. Also the case $p = 2$ and $q = 1$ is covered by [39, Theorems 4.2 and 4.3].

4.2.2 Approximation of multivariate Lipschitz functions

We study the problem of L_q-approximation of functions from the class

$$F_d = \{f : [0,1]^d \to \mathbb{R} : |f(x) - f(y)| \le \operatorname{dist}(x,y)\ \forall x,y \in [0,1]^d\}. \tag{4.4}$$

Here, dist denotes the maximum metric on the d-torus, i. e.,

$$\operatorname{dist}(x,y) = \min_{k \in \mathbb{Z}^d} \|x + k - y\|_\infty \quad \text{for} \quad x,y \in [0,1]^d.$$

We consider standard information of the form

$$N_n(f) = (f(x_1), \dots, f(x_n)), \quad x_1, \dots, x_n \in [0,1]^d.$$

An optimal algorithm based on N_n works as follows:

Algorithm. For N_n as above and $f \in F_d$, we define $A_n(f) = (f^+ + f^-)/2$, where

$$f^+(x) = \min_{i \le n}(f(x_i) + \operatorname{dist}(x, x_i)) \quad \text{and} \quad f^-(x) = \max_{i \le n}(f(x_i) - \operatorname{dist}(x, x_i)).$$

Note that f^+ and f^- are the maximal and the minimal function in F_d that interpolate f at the nodes $x_1, \dots, x_n$. We have the following relations that we prove for the sake of completeness, see e. g. [30, 37]:

Lemma 4.4.

$$r(N_n) = \sup_{f\in F_d} \|f - A_n(f)\|_q = \sup_{\substack{f\in F_d\\ N_n(f)=0}} \|f\|_q.$$

Proof. Clearly A_n is of the form $\phi \circ N_n$ for some mapping $\phi : \mathbb{R}^n \to L_q$. The definition of the radius yields

$$\sup_{f\in F_d} \|f - A_n(f)\|_q \ge r(N_n) \ge \sup_{\substack{f\in F_d\\ N_n(f)=0}} \|f\|_q.$$

On the other hand, any $f \in F_d$ satisfies the pointwise estimate

$$|f - A_n(f)| \le \frac{f^+ - f^-}{2}. \tag{4.5}$$

This implies

$$\sup_{\substack{f\in F_d\\ N_n(f)=0}} \|f\|_q \ge \sup_{f\in F_d} \|f - A_n(f)\|_q,$$

since the right-hand side of (4.5) is an element of F_d that vanishes on $\{x_1, \dots, x_n\}$. Altogether, we obtain the claimed identities. □

It is also well known that optimal information is given by function values on a regular grid if $n = m^d$ for some $m \in \mathbb{N}$. The radius of optimal information N_n^* satisfies

$$r(N_n^*) \asymp n^{-1/d}$$

for all $1 \le q \le \infty$. This follows from the upper bound on the complexity of uniform approximation as studied in [35], and the lower bound on the complexity of numerical integration as studied in [36]. Using the technique of proof of the latter, we even obtain a precise formula for the radius of optimal information.

Proposition 4.5. *Let $n = m^d$ for some $m \in \mathbb{N}$. Then*

$$r(N_n^*) = \begin{cases} \frac{1}{2}\left(\frac{d}{d+q}\right)^{1/q} n^{-1/d} & \text{if } 1 \le q < \infty,\\ \frac{1}{2}\, n^{-1/d} & \text{if } q = \infty. \end{cases}$$

Proof. Let N_n be as above, and let $P_n = \{x_1, \dots, x_n\}$. Note that the function

$$\operatorname{dist}(\cdot, P_n)\colon [0,1]^d \to \mathbb{R}, \quad \operatorname{dist}(x, P_n) = \min_{y\in P_n} \operatorname{dist}(x, y)$$

is contained in F_d and vanishes on P_n. On the other hand, every other function $f \in F_d$ that vanishes on P_n must satisfy

$$|f(x)| \le \operatorname{dist}(x, P_n) \quad \text{for all } x \in [0,1]^d.$$

This yields

$$r(N_n) = \sup_{\substack{f \in F_d \\ N_n(f)=0}} \|f\|_q = \|\mathrm{dist}(\cdot, P_n)\|_q.$$

Let us first consider the case $q = \infty$. Since the volume of the union of the balls $B_r^\infty(x)$ over $x \in P_n$ is smaller than 1 for all $r < 1/(2m)$, there must be some $x \in [0,1]^d$ with $\mathrm{dist}(x, P_n) \geq r$. Therefore,

$$\|\mathrm{dist}(\cdot, P_n)\|_\infty \geq \frac{1}{2m},$$

where equality holds when $P_n = \{i/m : 0 \leq i < m\}^d$. Let us turn to the case $q < \infty$. Let λ^d denote Lebesgue measure on $[0,1]^d$. Then

$$r(N_n)^q = \int_{[0,1]^d} \mathrm{dist}(x, P_n)^q \,\mathrm{d}x = \int_0^\infty \lambda^d(\mathrm{dist}(x, P_n)^q \geq t)\mathrm{d}t.$$

Note that

$$\lambda^d(\mathrm{dist}(x, P_n)^q \geq t) = 1 - \lambda^d(\mathrm{dist}(x, P_n)^q < t) \geq 1 - n \cdot 2^d t^{d/q},$$

where equality holds if the sets $B_{t^{1/q}}^\infty(y)$ for $y \in P_n$ are pairwise disjoint. Hence,

$$\begin{aligned} r(N_n)^q &\geq \int_0^{(1/2m)^q} \lambda^d(\mathrm{dist}(x, P_n)^q \geq t)\mathrm{d}t \\ &\geq \frac{1}{(2m)^q} - 2^d n \int_0^{(1/2m)^q} t^{d/q}\mathrm{d}t = \frac{1}{(2m)^q}\frac{d}{d+q} \end{aligned}$$

with equality for $P_n = \{i/m : 0 \leq i < m\}^d$. This proves the statement. □

In the following, we study the expected radius of random information

$$N_n(f) = (f(x_1), \ldots, f(x_n)),$$

where the points $x_1, \ldots, x_n$ are independent and uniformly distributed in $[0,1]^d$. If q is finite, the qth moment of the radius at zero can be computed precisely.

Proposition 4.6. *Consider L_q-approximation of functions from F_d defined in (4.4) with $1 \leq q < \infty$, using function values at uniformly distributed points in $[0,1]^d$. Then*

$$\mathbb{E}[r(N_n)^q] = \frac{1}{2^q}\frac{n!}{(q/d+1)\cdots(q/d+n)}.$$

In particular, we have

$$\lim_{n\to\infty} n^{1/d}\sqrt[q]{\mathbb{E}[r(N_n)^q]} = \frac{1}{2}\sqrt[q]{\Gamma(q/d+1)}.$$

Proof. Let $P_n = \{x_1, \dots, x_n\}$. Recall that

$$r(N_n)^q = \int_{[0,1]^d} \operatorname{dist}(x, P_n)^q \, dx.$$

Note that $\operatorname{dist}(\cdot,\cdot)\colon [0,1]^d \times [0,1]^{nd} \to \mathbb{R}$ is continuous, and thus Borel measurable. Using Tonelli's theorem, we obtain

$$\mathbb{E}[r(N_n)^q] = \int_{[0,1]^d} \mathbb{E}[\operatorname{dist}(x, P_n)^q] \, dx.$$

We will show that the integrand of the latter integral is constant. Let us fix $x \in [0,1]^d$ and note that $\operatorname{dist}(x, P_n) \in [0, 1/2]$. For any $t \in [0, 1/2]$, we have

$$\operatorname{dist}(x, P_n) \geq t \iff \forall i \in \{1, \dots, n\} : x_i \notin B_t^{\infty}(x).$$

Thus,

$$\mathbb{P}[\operatorname{dist}(x, P_n) \geq t] = \left(1 - (2t)^d\right)^n.$$

The substitution $s = 1 - (2t^{1/q})^d$ and integration by parts yield

$$\begin{aligned}
\mathbb{E}[\operatorname{dist}(x, P_n)^q] &= \int_0^{2^{-q}} \mathbb{P}[\operatorname{dist}(x, P_n)^q \geq t] \, dt = \int_0^{2^{-q}} \left(1 - \left(2t^{1/q}\right)^d\right)^n dt \\
&= \frac{q/d}{2^q} \int_0^1 s^n (1-s)^{q/d-1} \, ds = \frac{1}{2^q} \frac{n!}{(q/d+1)\cdots(q/d+n)},
\end{aligned}$$

which implies the statement of the theorem. □

We now turn to the case of uniform approximation, i. e., $q = \infty$. In this case, the expected radius of information is closely related to the so called coupon collector's problem. This is the question for the random number τ_ℓ of coupons that a coupon collector has to collect to obtain a complete set of ℓ distinct coupons. The following facts on the distribution of τ_ℓ are well known, see [28]. Here, $H_\ell = \sum_{k=1}^{\ell} 1/k$ is the ℓth harmonic number. Note that $H_\ell \sim \log \ell$ as $\ell \to \infty$.

Lemma 4.7. *Let $(Y_i)_{i=1}^{\infty}$ be a sequence of random variables that are uniformly distributed in the set $\{1, \dots, \ell\}$, and let*

$$\tau_\ell = \min\{n \in \mathbb{N} : \{Y_1, \dots, Y_n\} = \{1, \dots, \ell\}\}.$$

Then

$$\mathbb{E}[\tau_\ell] = \ell H_\ell \quad \textit{and} \quad \operatorname{Var}[\tau_\ell] \leq \ell^2 \sum_{k=1}^{\ell} 1/k^2$$

and for any $c \in (0,\infty)$,

$$\mathbb{P}[\tau_\ell > \lceil c\,\ell \log \ell \rceil] \le \ell^{-c+1}.$$

This leads to the following estimates of the expected radius for $q = \infty$.

Proposition 4.8. *Consider* L_∞*-approximation of functions from* F_d *defined in* (4.4). *For* $n \in \mathbb{N}$, *let*

$$m_1 = \min\{m \in \mathbb{N} : m^d (H_{m^d} - 2) \ge n\},$$
$$m_2 = \max\{m \in \mathbb{N} : 2m^d \log(m^d) \le n\}.$$

Then

$$\frac{1}{4m_1} \le \mathbb{E}[r(N_n)] \le \frac{2}{m_2}.$$

Proof. Let $m \in \mathbb{N}$, and decompose $[0,1]^d$ into $\ell = m^d$ boxes

$$\prod_{i=1}^{d} \Big[\frac{k_i - 1}{m}, \frac{k_i}{m}\Big), \quad k_1, \dots, k_d \in \{1, 2, \dots, m\}$$

of equal volume. Again, let $P_n = \{x_1, \dots, x_n\}$. Recall that the radius of the information $N_n(f) = (f(x_1), \dots, f(x_n))$ is given by

$$r(N_n) = \max_{x \in [0,1]^d} \operatorname{dist}(x, P_n).$$

Therefore, $r(N_n)$ is bounded above by $1/m$ if every box contains a point of P_n, and bounded below by $1/(2m)$ if one of the boxes does not contain a point of P_n. Let A be the event that every box contains a point. Note that the number of random points x_i that it takes to hit all the boxes follows the distribution of the coupon collector's variable τ_ℓ as defined in Lemma 4.7. For the upper bound, we choose $m = m_2$. Lemma 4.7 yields

$$\mathbb{P}[A^c] = \mathbb{P}[\tau_\ell > n] \le 1/\ell,$$

and hence

$$\mathbb{E}[r(N_n)] \le \mathbb{P}[A] \cdot \frac{1}{m} + \mathbb{P}[A^c] \cdot 1 \le \frac{2}{m}.$$

For the lower bound, we choose $m = m_1$. Chebyshev's inequality yields

$$\mathbb{P}[A] = \mathbb{P}[\tau_\ell \le n] \le \mathbb{P}[\tau_\ell \le \ell H_\ell - 2\ell] \le \frac{\operatorname{Var}[\tau_\ell]}{4\ell^2} \le \frac{1}{2}.$$

We obtain

$$\mathbb{E}[r(N_n)] \ge \mathbb{P}[A^c]\,\frac{1}{2m} \ge \frac{1}{4m},$$

as it was to be proven. □

We arrive at the following result on the power of random information proved in [25]. Note that the case $q = \infty$ is already known from [2], where the authors study the uniform approximation of functions on $[0,1]^d$ with bounded r^{th} derivative.

Theorem 4.9 (Krieg). *Consider L_q-approximation of functions from F_d defined in* (4.4) *with $1 \le q \le \infty$, using function values at uniformly distributed points in $[0,1]^d$. Then*

$$\mathbb{E}[r(N_n)] \asymp \begin{cases} r(N_n^*) \asymp n^{-1/d} & \text{if } q < \infty \\ r(N_{n/\log n}^*) \asymp (\frac{n}{\log n})^{-1/d} & \text{if } q = \infty. \end{cases}$$

Proof. The statements on the rate of the optimal radius of information follow from Proposition 4.5. The upper bound on the expected radius in the first case follows from Proposition 4.6, since the expected radius is bounded above by its qth moment. The lower bound is trivial. The upper and lower bound on the expected radius in the second case follow from Proposition 4.8, since both m_1^d and m_2^d are of order $n/\log n$. □

Thus, in the sense of order of convergence, random information is as good as optimal information for the problem of L_q-approximation on F_d if $q < \infty$. If $q = \infty$, we loose a logarithmic factor.

Remark 4.10 (Modifications of F_d). The rates of convergence of the expected and the optimal radius do not change if we replace the maximum metric on the torus by some equivalent metric. The same holds true if we change the Lipschitz constant, or if we switch to the nonperiodic setting.

4.2.3 Classical Sobolev spaces

We study the general L_q-approximation problem for Sobolev spaces,

$$\text{APP}: W_p^s([0,1]^d) \to L_q([0,1]^d),$$

given by $\text{APP}(f) = f$ with standard information. A typical choice for the norm in this space is

$$\|f\|_{W_p^s([0,1]^d)} = \Big(\sum_{\beta \in \mathbb{N}_0^d : \|\beta\|_1 \le s} \|D^\beta f\|_p^p \Big)^{1/p}.$$

It is known that the radius of optimal information N_n^* satisfies

$$r(N_n^*) \asymp n^{-s/d+(1/p-1/q)_+}, \tag{4.6}$$

where we assume that $s > d/p$ holds, to ensure that the functions are continuous. This can be found in [29] for general Lipschitz domains and was known earlier for special

cases, such as the cube, see [29, Remark 24]. Here, we ask for the expected radius of random information N_n given by n independent and uniformly distributed points in $[0,1]^d$. We conjecture that random information is as good as optimal information, provided that $p > q$; whereas in the case $p \leq q$, there is a loss of a logarithmic factor. This conjecture is true in the following special cases:
- Section 4.2.1 covers univariate Sobolev spaces of smoothness 1, i. e., $s = d = 1$.
- Section 4.2.2 covers Lipschitz functions, i. e., $s = 1$ and $p = \infty$.
- The paper [2] covers uniform approximation on Hölder classes, i. e., $p = q = \infty$.

In the general case, we are only able to prove the second part of our conjecture.

Theorem 4.11. *Consider L_q-approximation of functions from $W_p^s([0,1]^d)$ with $1 \leq p \leq q \leq \infty$, using function values at uniformly distributed points in $[0,1]^d$. Then*

$$\mathbb{E}[r(N_n)] \asymp r(N^*_{n/\log n}) \asymp \left(\frac{n}{\log n}\right)^{-s/d+1/p-1/q}.$$

We use the following criterion for optimal point sets, which can be found in [29]:

Lemma 4.12. *Consider L_q-approximation of functions from $W_p^s([0,1]^d)$ with $1 \leq p, q \leq \infty$. For all $n \in \mathbb{N}$, let $P_n \subset [0,1]^d$ be a point set of cardinality n,*

$$d_n = \min_{x,y \in P_n} \|x - y\|_\infty, \quad \text{and} \quad r_n = \max_{y \in [0,1]^d} \min_{x \in P_n} \|x - y\|_\infty.$$

If $r_n \preccurlyeq d_n$, then the information N_n with nodes P_n satisfies $r(N_n) \asymp r(N^*_n)$.

Proof of Theorem 4.11. Let $\alpha = s/d - 1/p + 1/q$. We start with the lower bound. Let $g : \mathbb{R}^d \to \mathbb{R}$ be a smooth function with support in the ℓ_∞-unit ball that is not identically zero. Using Lemma 4.7, we obtain with probability 1/2 that $[0,1]^d$ contains an ℓ_∞-ball of radius

$$\tau_n \asymp \left(\frac{\log n}{n}\right)^{1/d},$$

not containing any of the points in P_n. If x_0 denotes the center of this ball, then the function $f \in W_p^s([0,1]^d)$, given by $f(x) = g(\tau_n^{-1}(x - x_0))$, has support in this ball and satisfies

$$\|f\|_{W_p^s([0,1]^d)} \asymp \tau_n^{-s+d/p} \quad \text{and} \quad \|f\|_q \asymp \tau_n^{d/q}.$$

This yields

$$\mathbb{E}[r(N_n)] \geq \frac{1}{2} \frac{\|f\|_q}{\|f\|_{W_p^s([0,1]^d)}} \asymp \tau_n^{s-d/p+d/q} \asymp \left(\frac{n}{\log n}\right)^{-\alpha}.$$

We turn to the upper bound. We choose $m \in \mathbb{N}_0$ maximal such that $\ell = (3m)^d$ satisfies

$$n \geq (\alpha + 1)\ell \log \ell.$$

We split the unit cube into $k = m^d$ subcubes of equal volume. Moreover, we split each of these into 3^d subcubes of equal volume. This gives us $\ell = (3m)^d$ small cubes. Let A be the event that we hit every small cube. Lemma 4.7 yields

$$\mathbb{P}[A^c] = \mathbb{P}[\tau_\ell > n] \leq \ell^{-\alpha} \asymp \left(\frac{n}{\log n}\right)^{-\alpha}. \tag{4.7}$$

Assume now that A takes place. We choose exactly one point out of the small cube in the center of every large cube to obtain a subset P_k of P_n with cardinality k. Then

$$r_k \leq \frac{2}{3m} \leq d_k.$$

Let N_k' be the information map with nodes P_k. Lemma 4.12 yields

$$r(N_n) \leq r(N_k') \preccurlyeq k^{-\alpha} \asymp \left(\frac{n}{\log n}\right)^{-\alpha}. \tag{4.8}$$

Together, (4.7) and (4.8) imply the statement. □

We note that the authors of [12] study the integration problem for very general Sobolev spaces on closed manifolds. One can obtain the optimal rate $r(N_n^*) \asymp n^{-s/d}$. The authors prove the upper bound

$$\mathbb{E}[r(N_n)] \preccurlyeq n^{-s/d} \log(n)^{s/d},$$

where the random points are chosen with respect to the probability measure that defines the integral to be approximated. The given proof works with the covering radius of random points, which is an important characteristic for $q = \infty$. For the integration problem or $q = 1$, we expect, however, that the additional log factor is needed only for $p = 1$.

Remark 4.13. Another important case is presented by tensor products of univariate Sobolev spaces. These spaces are usually called Sobolev spaces with (dominating) mixed smoothness or, in the periodic case, sometimes Korobov spaces. One of the main features is that the optimal (main) orders of convergence are independent of the dimension, only the logarithmic terms depend on the dimension. We refer to [11] for a survey. There exist upper bounds for the numbers $\mathbb{E}[r(N_n)]$, but these bounds are weaker than for isotropic Sobolev spaces. One reason is that Lemma 4.12 does not hold for these spaces, and the geometrical structure of good sampling points is not known. See [1, 17, 20, 22, 30, 32, 33] for upper bounds on $\mathbb{E}[r(N_n)]$ in various cases.

Remark 4.14. Recovery of functions of many variables usually suffers from the curse of dimensionality, even for smooth C^∞ functions, see [31]. To avoid the curse, one can impose structural properties, such as the sparsity with respect to the Fourier coefficients or another orthonormal system. Then one can find an approximation of the function with relatively few random function values, see [6, 8]. Another way to study structural properties is to define weighted norms, as was first done in [34], see [30, 32, 33] for a survey.

4.3 Uniformly distributed linear information

In this section, we study *linear information* consisting of arbitrary continuous linear functionals. We are going to present two recovery problems, for which we compare random information with optimal information. The first is a classical result due to Kashin, Garnaev, and Gluskin [14, 24] on the recovery of ℓ_1^m-vectors in the Euclidean distance, whereas the second are novel results of the authors on the recovery of vectors in an m-dimensional ellipsoid in the Euclidean distance [18].

4.3.1 Recovery of ℓ_1^m-vectors in the Euclidean norm

We study the problem of recovery of vectors in the unit ball of ℓ_1^m with general linear information and error measured in the Euclidean norm. This is the approximation problem with $F = \ell_1^m$ and $G = \ell_2^m$, i. e., S is the identity map $\ell_1^m \hookrightarrow \ell_2^m$. The minimal worst-case error, which is the minimal radius of information achievable with general linear information, is closely connected to the notion of Gelfand numbers or Gelfand widths. Gelfand numbers and their dual counterpart, Kolmogorov numbers, are fundamental concepts in classical and modern approximation and complexity theory.

For $n \in \mathbb{N}$, the nth Gelfand number of a linear bounded operator $S : F \to G$ between normed spaces F and G can be defined as

$$c_n(S) = \inf_{\substack{F_n \subseteq F \\ \operatorname{codim}(F_n) \le n}} \sup_{\|f\|_F \le 1, f \in F_n} \|S(f)\|_G.$$

If the subspace F_n of F is the kernel of a linear information map $N_n : S \to \mathbb{R}^n$, then

$$r(N_n, 0) = \sup_{\|f\|_F \le 1, f \in F_n} \|S(f)\|_G$$

is the local radius of zero information. It is not hard to see that

$$r(N_n, 0) \le r(N_n) \le 2r(N_n, 0).$$

Since every subspace F_n of codimension at most n is the kernel of a linear information map $N_n : S \to \mathbb{R}^n$, we have the well-known relation

$$c_n(S) \le r(N_n^*) \le 2c_n(S)$$

between the radius of the optimal information N_n^* for the linear problem $S : F \to G$ and the Gelfand numbers $c_n(S)$.

The systematic study of the Gelfand numbers of identity maps $\ell_p^m \hookrightarrow \ell_q^m$ has a long tradition, and is the basis for the study of Gelfand numbers of embeddings of Sobolev spaces and, therefore, also for the complexity of L_q-approximation of functions from Sobolev spaces. We want to discuss here the case $p = 1$ and $q = 2$, which was one of the difficult cases. Its solution is an example of the power of random information. Whereas lower bounds had been obtained before, the breakthrough regarding the upper bounds was made using random approximation, a groundbreaking method having its origin in the work of Kashin [24]. The problem was finally settled in the series of papers by Kashin [23, 24], Gluskin [15, 16], and Garnaev and Gluskin [14].

Let us consider now the particular case of the identity $\ell_1^m \hookrightarrow \ell_2^m$, which is of importance in information-based complexity, since it also demonstrates that nonlinear recovery algorithms can be much better than linear ones. Although the results in the literature are usually formulated in the language of Gelfand numbers, we discuss the results directly for the radius of information.

Recall that we now allow arbitrary linear information $N_n : \ell_1^m \to \mathbb{R}^n$. The random information we consider is Gaussian information, where $N_n = N_{n,m}$ is given by an $n \times m$-matrix with independent standard normal entries. The kernels of these maps are distributed according to the Haar measure on the Grassmannian manifold of n-codimensional subspaces in $\mathbb{R}^m$. In this sense, we can speak about uniformly distributed linear information.

In [24, Theorem 1], Kashin introduced his fundamentally new approach via random subspaces and obtained that

$$\mathbb{E}\, r(N_{n,m}) \preccurlyeq \frac{1}{\sqrt{n}}\left(1 + \log \frac{m}{n}\right)^{3/2}.$$

Through a refinement of Kashin's arguments, Garnaev and Gluskin [14] were able to decrease the power 3/2 to a square root, while at the same time providing the matching lower bounds. Altogether, we have the following result:

Theorem 4.15 (Kashin, Garnaev, Gluskin). *Consider the recovery problem of vectors from ℓ_1^m in the Euclidean norm and Gaussian information. Then*

$$\mathbb{E}\, r(N_{n,m}) \asymp r(N_{n,m}^*) \asymp \min\left\{1, \sqrt{\frac{\log(1 + \frac{m}{n})}{n}}\right\}.$$

We want to sketch the proof of the upper bound

$$\mathbb{E}\, r(N_{n,m}) \preccurlyeq \min\left\{1, \sqrt{\frac{\log(1+\frac{m}{n})}{n}}\right\}$$

of Theorem 4.15 and demonstrate the power of the probabilistic method of Kashin, Garnaev, and Gluskin.

Step 1. As already explained, one considers a Gaussian random matrix $G = (g_{ij})_{i,j=1}^{n,m}$ with independent and identically distributed standard normal entries and defines the random information mapping

$$N_n : \mathbb{R}^m \to \mathbb{R}^n, \quad N_n(x) = Gx.$$

Step 2. One defines the mapping ϕ to be

$$\phi : \mathbb{R}^n \to \mathbb{R}^m, \quad y \mapsto \operatorname*{argmin}_{x\in\mathbb{R}^m,\, N_n(x)=y} \|x\|_1$$

and considers the corresponding random algorithm $A_n = \phi \circ N_n$.

Step 3. One can now prove that with high probability, the error of the random algorithm is small for any $x \in \mathbb{B}_1^m$. More precisely, if $\delta \in (0, \infty)$, then taking

$$n \asymp \frac{\log m}{\varepsilon^2} + \log \frac{2}{\delta}$$

yields a worst-case error

$$\sup_{x\in\mathbb{B}_1^m} \|x - A_n(x)\|_2 \le \varepsilon$$

for the random algorithm A_n with probability at least $1 - \delta$.

As a matter of fact, one actually shows that most linear information mappings $N_n : \mathbb{R}^m \to \mathbb{R}^n$ lead to a small error, and so random information is essentially as good as optimal information. We want to emphasize that no explicit construction for such a mapping is known. So the optimal information is not accessible in this case, and one has to settle for random information. We also want to mention that linear algorithms corresponding to linear widths or approximation numbers are much worse than nonlinear algorithms for this recovery problem.

Let us mention that Garnaev and Gluskin actually obtained sharp bounds in the more general setting of recovery of vectors from ℓ_1^m in the ℓ_q^m-norm or Gelfand numbers of the identity $\ell_1^m \hookrightarrow \ell_q^m$. If the random information $N_{n,m}$ is again Gaussian information, and $1 < q \le 2$, then

$$\mathbb{E}\, r(N_{n,m}) \asymp r(N_{n,m}^*) \asymp \min\left\{1, \left(\frac{\log(1+\frac{m}{n})}{n}\right)^{1-1/q}\right\}.$$

Later, Gelfand widths have also attracted quite some attention in the area of compressive sensing over the last two decades, see, e. g., [3, 4, 7, 10]. The goal of compressive sensing is to recover compressible vectors $x \in \mathbb{R}^m$ (those which are close to sparse vectors with only few nonzero coordinates) from n pieces of incomplete linear information. In fact, vectors in the unit ball of ℓ_p^m with $0 < p \le 1$ serve as a good model for sparse vectors. In this context, also the recovery of vectors in ℓ_p^m with $0 < p \le 1$ or the Gelfand numbers of the identity $\ell_p^m \hookrightarrow \ell_q^m$ were studied. Donoho [10] and Foucart, Pajor, Rauhut, and T. Ullrich [13, Theorem 1.1] extended the Garnaev–Gluskin result also to the case $0 < p \le 1$ and $p < q \le 2$, and proved

$$\mathbb{E}\, r(N_{n,m}) \asymp r(N_{n,m}^*) \asymp \min\left\{1, \left(\frac{\log(1+\frac{m}{n})}{n}\right)^{1/p-1/q}\right\}.$$

Again, it is not known how to explicitly construct measurement maps performing almost as good as $N_{n,m}^*$.

A further problem in compressed sensing is the matrix recovery problem. The analogon to the recovery problem of vectors from ℓ_p^m in the ℓ_q^m-norm is the recovery problem of matrices from Schatten p-classes in the Schatten q-norm. The particular case $q = 2$ is the Frobenius norm. This problem was studied in [5, 21].

4.3.2 Recovery of vectors from an ellipsoid

We continue with another recovery problem and recent results of the authors concerning the quality of random information for ℓ_2-approximation, see [18].

Let F be a centered ellipsoid in $\mathbb{R}^m$ with semi-axes $\sigma_1 \ge \cdots \ge \sigma_m$. Clearly, F can be seen as the unit ball of some Hilbert space. We want to recover $x \in F$ from the data $N_n(x) \in \mathbb{R}^n$ with linear information mapping $N_n \in \mathbb{R}^{n\times m}$, and measure the error in the Euclidean distance. Whereas the power of the information mapping is

$$r(N_n) = \inf_{\phi:\mathbb{R}^n\to\mathbb{R}^m} \sup_{x\in F} \|\phi(N_n(x)) - x\|_2,$$

it is known that, for linear problems in Hilbert spaces, the worst data is the zero data, and so

$$r(N_n) = \sup_{x\in F\cap E_n} \|x\|_2,$$

where E_n is the kernel of N_n, see [30, 37]. In this case, the power or radius of optimal information is given by $r(N_n^*) = \sigma_{n+1}$. The question is now how good random information N_n is for the ℓ_2-recovery problem, when random information is provided by a Gaussian random matrix $G = (g_{ij})_{i,j=1}^{n,m}$? Is it comparable to optimal information?

As is explained in [18], rephrased in geometric terms, one is interested in the circumradius of the intersection of a centered ellipsoid F in $\mathbb{R}^m$ with a random subspace

E_n of codimension n. Whereas the maximal radius is the length of the largest semi-axis σ_1, the minimal radius is the length of the $(n+1)^{\text{st}}$ largest semi-axis σ_{n+1}. But how large is the radius of a typical intersection? Is it comparable to the minimal or the maximal radius, or does it behave completely different?

Using various probabilistic tools, such as exponential estimates for sums of chi-squared random variables, Gordon's min-max theorem for Gaussian processes, or estimates for the extreme singular values of (structured) Gaussian matrices, the authors obtained upper and lower bounds on the radius of Gaussian information for the ℓ_2-approximation problem, which hold with overwhelming probability. In many cases, these bounds are sharp up to absolute constants and show that random information is comparable to optimal information.

Let us only present one case, which is of particular importance, where the semi-axes behave like the singular values of Sobolev embeddings [18, Corollary].

Theorem 4.16 (Hinrichs, Krieg, Novak, Prochno, Ullrich). *For $k \in \mathbb{N}$, let*

$$\sigma_k \asymp k^{-\alpha} \ln^{-\beta}(k+1),$$

where $\alpha \in (0,\infty)$ and $\beta \in \mathbb{R}$. Then

$$\mathbb{E}[r(N_n)] \asymp \begin{cases} \sigma_1 & \text{for } n < c_m, \text{ if } \alpha < 1/2 \text{ or } \beta \le \alpha = 1/2, \\ \sigma_{n+1}\sqrt{\ln(n+1)} & \text{for } n < \sqrt{m}, \text{ if } \beta > \alpha = 1/2, \\ \sigma_{n+1} & \text{for } n < m, \text{ if } \alpha > 1/2, \end{cases}$$

with

$$c_m = \begin{cases} c\, m^{1-2\alpha} \ln^{-\max\{2\beta,0\}} m & \text{for } \alpha < 1/2, \\ c \ln^{1-\max\{2\beta,0\}} m & \text{for } \beta < \alpha = 1/2, \\ c \ln\ln m & \text{for } \beta = \alpha = 1/2, \end{cases}$$

where $c \in (0,\infty)$ is an absolute constant.

This means that random information is just as good as optimal information if the singular values decay with a polynomial rate greater than 1/2. Then, in geometric terms, the size of a typical intersection ellipsoid is comparable to the size of the smallest intersection. On the other hand, if the singular values decay too slowly, random information is rather useless, which means that a typical intersection ellipsoid is almost as large as the largest.

The ellipsoid recovery problem can be considered in an infinite dimensional setting as well, even though our geometric intuition might fail. In this case, where $m = \infty$, we consider the ellipsoid

$$F = \left\{ x \in \ell_2 \colon \sum_{j\in\mathbb{N}} \frac{x_j^2}{\sigma_j^2} \le 1 \right\}$$

and the matrix $(g_{ij})_{1\le i\le n, j\in\mathbb{N}}$ with independent standard Gaussian entries as random information mapping N_n.

As a consequence of the probabilistic estimates derived in [18, Theorems 3 and 4], one obtains an ℓ_2-dichotomy showing that random information is useful if and only if $\sigma \in \ell_2$ [18, Corollary 5].

Theorem 4.17 (Hinrichs, Krieg, Novak, Prochno, Ullrich). *If $\sigma \notin \ell_2$, then $r(N_n) = \sigma_1$ holds almost surely for all $n \in \mathbb{N}$. If $\sigma \in \ell_2$, then*

$$\lim_{n\to\infty} \sqrt{n}\,\mathbb{E}[r(N_n)] = 0.$$

Remark 4.18. Such an ℓ_2-dichotomy is known from a related problem. There F is the unit ball of a reproducing kernel Hilbert space H, i. e., $H \subset L_2(D)$ consists of functions on a common domain D, and the evaluation functional $f \mapsto f(x)$ is a bounded operator on H for every $x \in D$. The optimal linear information N_n for the L_2-approximation problem is given by the singular value decomposition and has radius σ_{n+1}. This information might be difficult to implement, and hence one might allow only information N_n of the form

$$N_n(f) = (f(x_1), \dots, f(x_n)), \quad x_1, \dots, x_n \in D.$$

The goal is to relate the power of function evaluations to the power of all continuous linear functionals, and one would like to prove that their power is roughly the same. Unfortunately, this is *not* true in general. When $\sigma \notin \ell_2$, the convergence of optimal algorithms that may only use function values can be arbitrarily slow [19]. The situation is much better if we assume that $\sigma \in \ell_2$. It was shown in [27] and [38] that function values are not much worse than general linear information. We refer to [33, Chapter 26] for a presentation of these results. These results were recently improved in [26], where it was shown that the polynomial order of convergence for function values and general linear information is the same.

Bibliography

[1] F. Bach. On the equivalence between kernel quadrature rules and random feature expansions. *J. Mach. Learn. Res.*, 18: Paper No. 21, 38, 2017.

[2] B. Bauer, L. Devroye, M. Kohler, A. Krzyżak, and H. Walk. Nonparametric estimation of a function from noiseless observations at random points. *J. Multivar. Anal.*, 160: 93–104, 2017.

[3] E. J. Candès, J. Romberg, and T. Tao. Robust uncertainty principles: exact signal reconstruction from highly incomplete frequency information. *IEEE Trans. Inf. Theory*, 52(2): 489–509, 2006.

[4] D. Chafaï, O. Guédon, G. Lecué, and A. Pajor. *Interactions between compressed sensing random matrices and high dimensional geometry*, volume 37 of *Panoramas et Synthèses [Panoramas and Syntheses]*. Société Mathématique de France, Paris, 2012.

[5] J. A. Chávez-Domínguez and D. Kutzarova. Stability of low-rank matrix recovery and its connections to Banach space geometry. *J. Math. Anal. Appl.*, 427(1): 320–335, 2015.

[6] A. Chkifa, N. Dexter, H. Tran, and C. G. Webster. Polynomial approximation via compressed sensing of high-dimensional functions on lower sets. *Math. Comput.*, 87(311): 1415–1450, 2018.
[7] A. Cohen, W. Dahmen, and R. DeVore. Compressed sensing and best k-term approximation. *J. Am. Math. Soc.*, 22(1): 211–231, 2009.
[8] A. Cohen, R. DeVore, S. Foucart, and H. Rauhut. Recovery of functions of many variables via compressive sensing. In *Proc. SampTA 2011*, Singapore, 2010.
[9] D. A. Darling. On a class of problems related to the random division of an interval. *Ann. Math. Stat.*, 24: 239–253, 1953.
[10] D. L. Donoho. Compressed sensing. *IEEE Trans. Inf. Theory*, 52(4): 1289–1306, 2006.
[11] D. Dung, V. T. Temlyakov, and T. Ullrich. *Hyperbolic Cross Approximation*. Advanced Courses in Mathematics – CRM Barcelona. Springer International Publishing, 2018.
[12] M. Ehler, M. Graef, and C. J. Oates. Optimal Monte Carlo integration on closed manifolds. *Stat. Comput.*, 29: 1203–1214, 2019.
[13] S. Foucart, A. Pajor, H. Rauhut, and T. Ullrich. The Gelfand widths of ℓ_p-balls for $0 < p \leq 1$. *J. Complex.*, 26(6): 629–640, 2010.
[14] A. Yu. Garnaev and E. D. Gluskin. The widths of a Euclidean ball. *Dokl. Akad. Nauk SSSR*, 277(5): 1048–1052, 1984. English translation: Soviet Math. Dokl. 30 (1984), no. 1, 200–204.
[15] E. D. Gluskin. On some finite-dimensional problems of width theory. *Physis—Riv. Internaz. Storia Sci.*, 23(2): 5–10, 124, 1981.
[16] E. D. Gluskin. Norms of random matrices and diameters of finite-dimensional sets. *Mat. Sb. (N. S.)*, 120(162)(2): 180–189, 286, 1983.
[17] S. Heinrich, E. Novak, G. W. Wasilkowski, and H. Woźniakowski. The inverse of the star-discrepancy depends linearly on the dimension. *Acta Arith.*, 96(3): 279–302, 2001.
[18] A. Hinrichs, D. Krieg, E. Novak, J. Prochno, and M. Ullrich. Random sections of ellipsoids and the power of random information. *arXiv e-prints*, page arXiv:1901.06639, Jan 2019.
[19] A. Hinrichs, E. Novak, and J. Vybíral. Linear information versus function evaluations for L_2-approximation. *J. Approx. Theory*, 153(1): 97–107, 2008.
[20] A. Hinrichs, J. Oettershagen, and M. Ullrich. Numerical integration of smooth functions using random point sets. In preparation.
[21] A. Hinrichs, J. Prochno, and J. Vybíral. Entropy numbers of embeddings of Schatten classes. *J. Funct. Anal.*, 273(10): 3241–3261, 2017.
[22] M. Kanagawa, B. K. Sriperumbudur, and K. Fukumizu. Convergence analysis of deterministic kernel-based quadrature rules in misspecified settings. *Found. Comput. Math.*, 20: 155–194, 2020.
[23] B. S. Kašin. Kolmogorov diameters of octahedra. *Dokl. Akad. Nauk SSSR*, 214: 1024–1026, 1974.
[24] B. S. Kašin. The widths of certain finite-dimensional sets and classes of smooth functions. *Izv. Akad. Nauk SSSR, Ser. Mat.*, 41(2): 334–351, 478, 1977. English translation: Math. USSR-Izv. 11 (1977), no. 2, 317–333 (1978).
[25] D. Krieg. *Algorithms and Complexity of some Multivariate Problems*. PhD thesis, Friedrich Schiller University Jena, 2019.
[26] D. Krieg and M. Ullrich. Function values are enough for L_2-approximation. *arXiv e-prints*, page arXiv:1905.02516, May 2019.
[27] F. Y. Kuo, G. W. Wasilkowski, and H. Woźniakowski. On the power of standard information for multivariate approximation in the worst case setting. *J. Approx. Theory*, 158(1): 97–125, 2009.
[28] D. A. Levin, Y. Peres, and E. L. Wilmer. *Markov chains and mixing times*. American Mathematical Society, Providence, RI, 2009. With a chapter by James G. Propp and David B. Wilson.

[29] E. Novak and H. Triebel. Function spaces in Lipschitz domains and optimal rates of convergence for sampling. *Constr. Approx.*, 23(3): 325–350, 2006.

[30] E. Novak and H. Woźniakowski. *Tractability of multivariate problems. Vol. 1: Linear information*, volume 6 of *EMS Tracts in Mathematics*. European Mathematical Society (EMS), Zürich, 2008.

[31] E. Novak and H. Woźniakowski. Approximation of infinitely differentiable multivariate functions is intractable. *J. Complex.*, 25(4): 398–404, 2009.

[32] E. Novak and H. Woźniakowski. *Tractability of multivariate problems. Volume II: Standard information for functionals*, volume 12 of *EMS Tracts in Mathematics*. European Mathematical Society (EMS), Zürich, 2010.

[33] E. Novak and H. Woźniakowski. *Tractability of multivariate problems. Volume III: Standard information for operators*, volume 18 of *EMS Tracts in Mathematics*. European Mathematical Society (EMS), Zürich, 2012.

[34] I. H. Sloan and H. Woźniakowski. When are quasi-Monte Carlo algorithms efficient for high-dimensional integrals? *J. Complex.*, 14(1): 1–33, 1998.

[35] A. G. Sukharev. An optimal method for the construction of the best uniform approximations for a certain class of functions. *Ž. Vyčisl. Mat. Mat. Fiz.*, 18(2): 302–313, 523, 1978. English translation: U. S. S. R. Comput. Math. Math. Phys. 18 (1978), no. 2, 21–31.

[36] A. G. Sukharev. Optimal formulas of numerical integration for some classes of functions of several variables. *Dokl. Akad. Nauk SSSR*, 246(2): 282–285, 1979. English translation: Soviet Math. Dokl. 20 (1979), no. 3, 472–475.

[37] J. F. Traub, G. W. Wasilkowski, and H. Woźniakowski. *Information-based complexity*. Computer Science and Scientific Computing. Academic Press, Inc., Boston, MA, 1988. With contributions by A. G. Werschulz and T. Boult.

[38] G. W. Wasilkowski and H. Woźniakowski. On the power of standard information for weighted approximation. *Found. Comput. Math.*, 1(4): 417–434, 2001.

[39] H. Weyhausen. *Expected Discrepancies*. PhD thesis, Friedrich Schiller University Jena, 2015.

C. J. Oates, J. Cockayne, D. Prangle, T. J. Sullivan, and M. Girolami

5 Optimality criteria for probabilistic numerical methods

Abstract: It is well understood that Bayesian decision theory and average case analysis are essentially identical. However, if one is interested in performing *uncertainty quantification* for a numerical task, it can be argued that standard approaches from the decision-theoretic framework are neither appropriate nor sufficient. Instead, we consider a particular optimality criterion from Bayesian experimental design and study its implied optimal information in the numerical context. This information is demonstrated to differ, in general, from the information that would be used in an average-case-optimal numerical method. The explicit connection to Bayesian experimental design suggests several distinct regimes, in which optimal probabilistic numerical methods can be developed.

Keywords: Average case analysis, Bayesian decision theory, experimental design

MSC 2010: 65-02, 62A01

5.1 Introduction

To fix notation, consider the task of approximating a quantity of interest $\phi: \mathcal{X} \to \Phi$ based on finite information provided through a map $y_e: \mathcal{X} \to \mathcal{Y}_e$. There may be several such maps with equivalent computational cost, and these are indexed by $e \in \mathcal{E}$. This abstract formulation covers most basic numerical tasks, including numerical integra-

Acknowledgement: The authors are grateful to an anonymous reviewer for their thoughtful comments, to Dave Woods for discussion of this work, as well as to the organizers and participants of the RICAM workshop on *Multivariate Algorithms and Information-Based Complexity*, for which this paper was prepared. CJO and MG were supported by the Lloyd's Register Foundation programme on Data-Centric engineering at the Alan Turing Institute, UK. TJS was supported by the Excellence Initiative of the German Research Foundation (DFG) through the Freie Universität Berlin. MG was supported by the EPSRC grants [EEP/P020720/1, EP/R018413/1, EP/R034710/1, EP/R004889/1] and a Royal Academy of Engineering Research Chair.

C. J. Oates, Newcastle University, Newcastle upon Tyne, UK; and Alan Turing Institute, London, UK, e-mail: chris.oates@ncl.ac.uk
J. Cockayne, University of Warwick, Coventry, UK, e-mail: j.cockayne@warwick.ac.uk
D. Prangle, Newcastle University, Newcastle upon Tyne, UK, e-mail: dennis.prangle@ncl.ac.uk
T. J. Sullivan, Freie Universität Berlin, Berlin, Germany; and Zuse Institute Berlin, Berlin, Germany, e-mail: sullivan@zib.de
M. Girolami, University of Cambridge, Cambridge, UK; and Alan Turing Institute, London, UK, e-mail: m.girolami@eng.cam.ac.uk

https://doi.org/10.1515/9783110635461-005

tion, numerical optimisation, and the numerical solution of a differential equation [64]. Thus, a *numerical method* is considered as a map $d_e\colon \mathcal{Y}_e \to \Phi$, and the set of all possible numerical methods is denoted $\mathcal{D}_e$.

The performance of a numerical method can be quantified in several ways, but a common scenario is for $\mathcal{X}$ and Φ to be normed spaces, in which case the *average-case error* of a numerical method d_e can be defined as

$$\mathrm{ACE}_p(e, d_e) := \left(\int \|d_e(y_e(x)) - \phi(x)\|_\Phi^p \, \mathrm{d}\pi_X(x) \right)^{1/p}$$

where $p \in [1, \infty)$, and π_X is a Borel probability distribution on $\mathcal{X}$, to be specified. Average-case analysis can be motivated by considering the state x to be the realisation of a random variable $X \sim \pi_X$, and then assessing the average performance of a numerical method [51].

In *information-based complexity* there is interest in optimal numerical methods, their associated optimal information, and the tractability of the numerical task itself. For instance, an *average-case-optimal numerical method* is defined as

$$d_e^* \in \operatorname*{arg\,min}_{d_e \in \mathcal{D}_e} \mathrm{ACE}_p(e, d_e), \tag{5.1}$$

and *average-case-optimal information* is defined as

$$e^* \in \operatorname*{arg\,min}_{e \in \mathcal{E}} \mathrm{ACE}_p(e, d_e^*). \tag{5.2}$$

See [51, 57, 58, 64]. Here and in the sequel, we simply assume all minima are realised, so (5.1) and (5.2) are presented as arg min rather than arg inf. In [31] and later [21], it was observed that an average-case-optimal numerical method is mathematically identical to a *Bayes rule* in Bayesian decision theory when π_X is interpreted as the *prior* distribution (see Section 5.2).

There has been recent interest in the development of *probabilistic numerical methods*, which treat the approximation of $\phi(x)$ as a statistical estimation task [29, 33]:

Definition 5.1 ([18]; Defn. 2.2). A *probabilistic numerical method* is a map $d_e\colon \mathcal{Y}_e \to \mathcal{P}_\Phi$, where $\mathcal{P}_\Phi$ is the set of all Borel probability distributions on Φ.

The motivation for probabilistic numerical methods is to provide formal uncertainty quantification for the quantity of interest. Recall that the push-forward $\phi_\# \pi$ of a distribution π is defined as $(\phi_\# \pi)(S) = \pi(\phi^{-1}(S))$ for all measurable S. Let $\pi_{Y|e} := (y_e)_\# \pi_X$, and denote the disintegration[1] of π_X along the map y_e by $\{\pi_{X|y,e}\}_{y \in \mathcal{Y}_e}$.

---

1 If all random variables admit densities with respect to a reference measure, then $\pi_{X|y,e}$ is the usual conditional distribution of X, given $y_e(X)$. A formal treatment of disintegration of measure is presented in [13].

Definition 5.2 ([18]; Defn. 2.5). A probabilistic numerical method $d_e : \mathcal{Y}_e \to \mathcal{P}_\Phi$ is said to be *Bayesian* with prior π_X if $d_e(y) = \phi_\# \pi_{X|y,e}$ for $\pi_{Y|e}$-almost all $y \in \mathcal{Y}_e$.

Thus, the output of a Bayesian probabilistic numerical method is a *posterior* distribution, providing uncertainty quantification for the quantity of interest. The foundations of Bayesian probabilistic numerical methods were established in [18] and Bayesian methods (in the strict sense of Defn. 5.2) have been studied for numerical integration [6, 32], global optimisation [28, 37, 38], ordinary [62] and partial [19, 49] differential equations, and linear algebra [3, 17, 27]. Moreover, these methods are starting to see practical application [14, 40, 50].

The distributional output of a Bayesian probabilistic numerical method could, of course, be reduced to a point estimator for the quantity of interest, i. e., a (traditional) numerical method. Standard calculations from Bayesian decision theory, for example, following [31], imply that the mean of $\phi_\# \pi_{X|y,e}$ is an average-case-optimal numerical method with $p = 2$ (see Section 5.2.2). Conversely, [21] illustrated that the trapezoidal rule for numerical integration is a Bayes rule, which is in turn an average-case-optimal numerical method, when π_X is the standard Weiner measure on $C(0,1)$. Despite their elegance, these connections rather overlook the main objective of a probabilistic numerical method, which is to provide formal uncertainty quantification for the quantity of interest. Indeed, to draw an analogy, it is well-understood that different strategies are required for the contrasting goals of parameter estimation and prediction in the statistical context [42]. In particular, the notion of optimal information put forward in average-case analysis (equivalently, Bayesian decision theory) is not necessarily an appropriate criteria on which to base the design of a Bayesian probabilistic numerical method.

The aim of this chapter is therefore twofold: First, we review connections between average-case-optimal numerical methods, average-case-optimal information, and approaches to Bayesian experimental design from the statistical literature. Second, we discuss how optimal information could be defined for a Bayesian probabilistic numerical method, where the goal is to perform uncertainty quantification, as opposed to direct estimation of a quantity of interest. In particular, we explore a particular criterion proposed in the recent work of [18] that has the advantage of being straightforward to approximate.

5.2 Bayesian decision theory

In this section, we present a succinct overview of statistical decision theory [4], recalling that this is mathematically identical to average-case analysis, as explained in [31]. Throughout, we use calligraphic font to represent (topological) state spaces, e. g., $\mathcal{X}$. Upper-case letters are used to denote (Borel) random variables, e. g., X. For convenience, we identify X with a random variable on $\mathcal{X}$, whose distribution is denoted as π_X. The elements of a state space are denoted with lower-case letters, e. g., $x \in \mathcal{X}$.

Consider again an index set $\mathcal{E}$. In what follows, we adopt the terminology of Bayesian decision theory and, in particular, refer to $e \in \mathcal{E}$ as an *experiment*. Our presentation will be more general than in Section 5.1 in three respects, which are all consequences of the broader context in which Bayesian decision-theoretic methods have been developed:

First, we associate with each experiment $e \in \mathcal{E}$ and each state $x \in \mathcal{X}$ a random variable $Y_e(x)$ taking values in $\mathcal{Y}_e$. The conditional distribution of this random variable is denoted $\pi_{Y|x,e}$. Often $Y_e(x) = y_e(x)$ with probability one, where $y_e(x)$ is a deterministic function of the state x, as in Section 5.1, in which case $\pi_{Y|x,e} = \delta(y_e(x))$ is an atomic measure centred on $y_e(x) \in \mathcal{Y}_e$. The more general formulation allows for the possibility that information about the state x is corrupted by noise. Let $\pi_{Y|e}$ be the distribution defined through marginalisation as

$$\pi_{Y|e}(S) = \iint 1_S(y)\, \mathrm{d}\pi_{Y|x,e}(y)\, \mathrm{d}\pi_X(x),$$

where 1_S is the indicator function for the measurable set $S \subset \mathcal{Y}_e$. The computational cost of each experiment is considered to be identical, but experiments differ in what information about x is provided in $Y_e(x)$.

Second, we formulate the goal of Bayesian decision theory as the selection of an *action* a from a specified set $\mathcal{A}$. Often, as in Section 5.1, the aim is to approximate $\phi(x)$, and the set of actions $\mathcal{A}$ is identical to Φ. The more general formulation allows for more compex policy and control strategies to be considered in the statistical context. Correspondingly, a *decision rule* $d_e \in \mathcal{D}_e$ is considered to be a function $d_e : \mathcal{Y}_e \to \mathcal{A}$.

Third, we consider an arbitrary *loss function* $\ell : \mathcal{X} \times \mathcal{A} \to \mathbb{R}$. This includes the case $\ell(x, a) = \|\phi(x) - a\|_\Phi^p$ from Section 5.1. The choice of the loss function should be informed by the reason why the experiment is being conducted and, in an estimation context, should reflect the quantities of interest. See Chapter 2 of [4] for further discussion of this point.

Example 5.1 (Numerical integration). To illustrate the notation, consider the numerical task of integrating a continuous function $x \in \mathcal{X} = C(0, 1)$. The aim here is to select an element $a \in \mathcal{A} = \mathbb{R}$, which represents an approximation to the quantity of interest $\phi(x) = \int_0^1 x(t)\, \mathrm{d}t$. Here, the experiment set $\mathcal{E}$ consists of vectors $e = [t_1, \dots, t_{n-1}]$, along with fixed endpoints $t_0 = 0$, $t_n = 1$, and $t_0 \le \dots \le t_n$, corresponding to maps $y_e(x) = [x(t_0), \dots, x(t_n)]$. One natural loss function is $\ell(x, a) = (\phi(x) - a)^2$. This problem has been well studied and will be used in an illustrative capacity in the sequel.

5.2.1 Bayes risk

The framework of statistical decision theory allows comparison of different decision rules based on their *Bayes risk*:

$$\mathrm{BR}(e, d_e) := \iint \ell(x, d_e(y))\, \mathrm{d}\pi_{Y|x,e}(y)\, \mathrm{d}\pi_X(x). \tag{5.3}$$

The Bayes risk quantifies the prior expected loss associated with a decision rule d_e, and therefore forms a natural criterion for the selection of a decision rule in the Bayesian context. We therefore define a *Bayes rule* for an experiment $e \in \mathcal{E}$ to be

$$d_e^* \in \arg\min_{d_e \in \mathcal{D}_e} \mathrm{BR}(e, d_e). \tag{5.4}$$

In a restricted context,[2] the definition of a Bayes rule is mathematically identical to the definition of an average-case-optimal numerical method in eq. (5.1). Similarly, we can define an optimal experiment,

$$e^* \in \mathcal{E}^*_{\mathrm{BDT}} := \arg\min_{e \in \mathcal{E}} \mathrm{BR}(e, d_e^*),$$

which is analogous to average-case-optimal information in eq. (5.2).

Example 5.2 (Numerical integration, continued). [54, 55] considered the standard Wiener measure π_X, a Gaussian measure on $C(0,1)$ characterized by the property that for all $t, t' \in [0,1]$ we have $\int x(t)\, \mathrm{d}\pi_X(x) = 0$ and $\int x(t)x(t')\, \mathrm{d}\pi_X(x) = \min(t, t')$. In that work, it was shown that for each experiment $e = [t_1, \dots, t_{n-1}]$ there exists a unique Bayes rule

$$d_e^* = \frac{1}{2} \sum_{i=1}^{n} (x(t_{i-1}) + x(t_i))(t_i - t_{i-1}),$$

which we recognise as the trapezoidal rule. Moreover, it was shown that there is a unique optimal experiment with $t_i = \frac{i}{n}$. Further contributions in this direction include [21, 33, 34, 41, 43, 44]. See [51] for a book-length treatment.

Remark 5.1 (Admissibility). It is important to note that other notions of optimality for decision rules, such as admissibility, need not coincide with the Bayesian notion of optimality. A decision rule $d_e \in \mathcal{D}_e$ is called *admissible* if there exists no $d_e' \in \mathcal{D}_e$ such that

$$\int \ell(x, d_e'(y))\, \mathrm{d}\pi_{Y|x,e}(y) \le \int \ell(x, d_e(y))\, \mathrm{d}\pi_{Y|x,e}(y)$$

---

2 Indeed, as highlighted in [31], eq. (5.3) is identical to $\mathrm{ACE}_p(e, d_e)^p$ when $\pi_{Y|x,e} = \delta(y_e(x))$, $\mathcal{A} = \Phi$, and $\ell(x, a) = \|\phi(x) - a\|_\Phi^p$.

for all $x \in \mathcal{X}$, with strict inequality for some $x \in \mathcal{X}$. The simplest illustration is the estimation of $x \in \mathbb{R}$ based on $Y|x \sim N(x, 1)$ with $\ell(x, x') = (x - x')^2$, where an admissible decision rule is $d(y) = y$, but this is not a Bayes rule for any proper prior on $\mathbb{R}$. (Throughout this contribution, $N(\mu, \Sigma)$ denotes a Gaussian distribution with mean μ and covariance Σ.) Some results on when Bayes rules are, and are not, admissible can be found in the references provided in Section 8.4 of [4]. The famous result of [61] demonstrated that, under certain conditions, all admissible decision rules are so-called *generalised* Bayes rules, a definition in which improper priors are permitted. For the purposes of uncertainty quantification, however, we wish to remain in the Bayesian framework and consider only Bayes rules that arise from, and can be understood in terms of, a probabilistic model.

5.2.2 Bayes acts

Following the definition of a Bayes rule, it is reasonable to ask what actions a Bayes rule would select. To this end, denote the set of *Bayes acts* as

$$\mathcal{A}_e^*(y_e) := \arg\min_{a \in \mathcal{A}} \int \ell(x, a) \, \mathrm{d}\pi_{X|y,e}(x).$$

The proof of the following result is provided in Appendix A.1:

Proposition 5.1. *A decision rule $d_e^* \in \mathcal{D}_e$ is a Bayes rule if and only if $d_e^*(y) \in \mathcal{A}_e^*(y)$ for $\pi_{Y|e}$-almost all $y \in \mathcal{Y}_e$.*

Often the set $\mathcal{D}_e$ will be infinite-dimensional, so that the search for a Bayes rule involves an optimisation problem, which is infinite-dimensional. However, the set $\mathcal{A}_e$ is often finite-dimensional. Hence, the action of a Bayes rule is something that can often be computed.

The next result is technical and will be needed later to deduce that the mean of $\pi_{X|y,e}$ is a Bayes act for a certain family of loss functions. Recall that a function $\phi\colon \mathcal{X} \to \mathbb{R}^m$ is called *coercive* if for all $c > 0$ there exists a compact set $K \subset \mathcal{X}$ such that $\|\phi(x)\|_2 \geq c$ for all $x \in \mathcal{X} \setminus K$. The proof of the following is provided in Appendix A.2:

Proposition 5.2. *Consider $\mathcal{A} = \mathcal{X} = \mathbb{R}^d$. Let $\ell(x, a) = \|\phi(x) - \phi(a)\|_2^2$, where $\phi\colon \mathcal{X} \to \mathbb{R}^m$, $m \in \mathbb{N}$. Assume that ϕ is twice continuously differentiable, that $\int \|\phi(x)\|_2 \, \mathrm{d}\pi_{X|y,e}(x)$ is finite, and that the matrix*

$$\left[\frac{\mathrm{d}\phi}{\mathrm{d}a}\right]_{i,j} = \frac{\partial}{\partial a_j} \phi_i(a) \tag{5.5}$$

has full row rank at all $a \in \mathcal{A}$. Then any Bayes act $a \in \mathcal{A}_e^*(y_e)$ satisfies

$$\phi(a) = \int \phi(x) \, \mathrm{d}\pi_{X|y,e}(x). \tag{5.6}$$

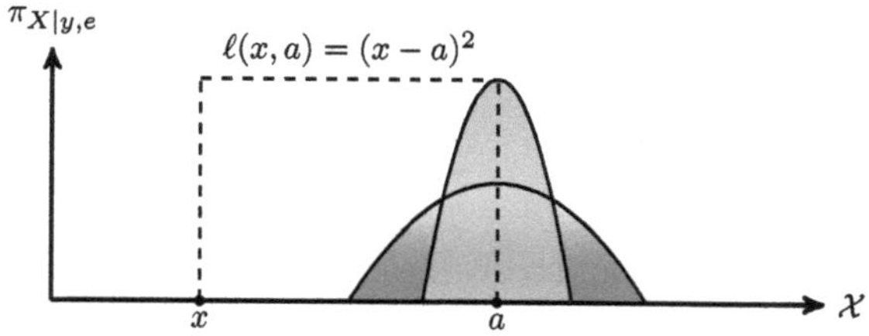

Figure 5.1: Two posteriors ($\pi_{X|y,e}$; blue and green densities) can lead to the same Bayes act (a; the posterior mean in the case of squared loss ℓ), but provide very different uncertainty quantification for a quantity of interest ($\phi(x)$).

Moreover, if there exists a unique solution to eq. (5.6) *and the function ϕ is coercive, then this solution is a Bayes act.*

For simplicity, we have presented this result in finite dimensions and for the standard Euclidean norm, but it can be naturally extended to infinite dimensions and to an arbitrary Hilbert space $\mathcal{X} = \mathcal{A} = \Phi$.

Example 5.3 (Linear regression). Let $\pi_X = N(\mu_0, \Sigma_0)$, $X \in \mathbb{R}^d$, and $\pi_{Y|x,e} = N(A_e x, \Sigma)$, $Y \in \mathbb{R}^n$, where the matrix Σ_0 is positive definite and the matrix $A_e \in \mathbb{R}^{n\times d}$ is determined by the choice of experiment $e \in \mathcal{E}$. Consider a loss $\ell(x, x') = (x - x')^\top \Lambda (x - x')$, where Λ is a positive semidefinite matrix with a square root $\Lambda^{\frac{1}{2}}$. Then a Bayes decision rule d_e^* is defined through the Bayes act(s) $a \in \mathbb{R}^d$ which, from Proposition 5.2, satisfy $\Lambda^{\frac{1}{2}} a = \Lambda^{\frac{1}{2}} \mu_{y,e}$, where $\pi_{X|y,e} = N(\mu_{y,e}, \Sigma_e)$, and $\Sigma_e = (A_e^\top A_e + \Sigma_0^{-1})^{-1}$, $\mu_{y,e} = \Sigma_e (A_e^\top y + \Sigma_0^{-1} \mu_0)$. If Λ is positive definite, then it also follows from Proposition 5.2 that $\mu_{y,e}$ is the unique Bayes act.

The explicit connection between Bayesian decision theory and average-case analysis [31] can be exploited to obtain information-based complexity results for Bayes rules and optimal experiments in the decision-theoretic context. However, we recall that the differing goals of parameter estimation and uncertainty quantification need not lead to the same notions of an optimal experiment. Indeed, two posteriors $\pi_{X|y,e}$ can lead to the same Bayes act, but provide very different uncertainty quantification for a quantity of interest; see Figure 5.1. To understand this point in detail, we turn to the more general framework of Bayesian experimental design in Section 5.3.

5.3 Bayesian experimental design

The term "Bayesian experimental design" has various historical usages, but in this work we are broadly consistent with the presentation of [12]. In this usage, Bayesian experimental design (BED) requires a *utility* function u to be specified, such that $u(e, x, y) \in \mathbb{R}$ represents the "usefulness" of the information $y \in \mathcal{Y}_e$ when the true state of nature is $x \in \mathcal{X}$. Armed with a utility function, we can construct a design criterion

$$\mathrm{BED}(e) := -\iint u(e,x,y)\,\mathrm{d}\pi_{Y|x,e}(y)\,\mathrm{d}\pi_X(x), \tag{5.7}$$

and the following notion of an optimal experiment is considered:

$$\mathcal{E}^*_{\mathrm{BED}} := \operatorname*{arg\,min}_{e\in\mathcal{E}} \mathrm{BED}(e).$$

See [16, 36, 39]. In general, optimal experimental designs will be analytically intractable, and in practice, sophisticated numerical approaches to approximate the integral in eq. (5.7), and to perform the global multivariate optimisation may be needed (e. g., [2, 39, 45, 46]). In an applied statistical context, the utility is something that itself may be elicited [63], and robustness to perturbation of the utility can be investigated [52].

The Bayesian experimental design framework is not concerned with selecting an action; it concerns only selection of an experiment. However, in terms of characterising an optimal experiment, the Bayesian experimental design framework is strictly more general than the Bayesian decision-theoretic approach, as we demonstrate next.

5.3.1 Design criteria via decision theory

First, we explain how the decision-theoretic approach can be recovered in the experimental design framework [36, pp. 19–20], [20]. This is achieved by considering utilities $u(e,x,y)$, which are independent of the state x, in particular

$$u(e,y) = -\int \ell(x', d^*_e(y))\,\mathrm{d}\pi_{X|y,e}(x'). \tag{5.8}$$

Under this choice, eq. (5.7) becomes

$$\begin{aligned}\mathrm{BED}(e) &= \iiint \ell(x', d^*_e(y))\,\mathrm{d}\pi_{X|y,e}(x')\,\mathrm{d}\pi_{Y|x,e}(y)\,\mathrm{d}\pi_X(x)\\ &= \iint \ell(x', d^*_e(y))\,\mathrm{d}\pi_{X|y,e}(x')\,\mathrm{d}\pi_{Y|e}(y)\\ &= \iint \ell(x', d^*_e(y))\,\mathrm{d}\pi_{Y|x',e}(y)\,\mathrm{d}\pi_X(x') \;=\; \mathrm{BR}(e, d^*_e),\end{aligned}$$

and so $\mathcal{E}^*_{\mathrm{BED}} = \mathcal{E}^*_{\mathrm{BDT}}$.

To overcome the fact that $\mathrm{BR}(e,d^*_e)$ is analytically intractable in general, a suite of approximations specific to the decision-theoretic context have been developed. Most approximations in the literature can be obtained by fixing a specific loss function ℓ, and then making a Gaussian approximation to the posterior; see [20] for a more theoretical treatment. In this section, we briefly present the approach and, for concreteness, we suppose $\mathcal{X} = \mathbb{R}^m$ so that x is a column vector. Therefore, we consider Gaussian approximations $\pi_{X|y,e} \approx N(\mu_{y,e}, \Sigma_e)$. Note that the covariance Σ_e is considered to be independent of the information y_e; a motivating example with this property was illustrated in Example 5.3. Then different choices of loss function ℓ lead to different approximations, the co-called *alphabet criteria*:

$$\mathrm{BR}(e, d_e^*) \approx \begin{cases} \operatorname{tr}(\Lambda\Sigma_e) & \text{A-optimal} \\ \det(\Lambda^{1/2}\Sigma_e\Lambda^{1/2}) & \text{D-optimal} \\ \vdots \end{cases}$$

for some positive semidefinite matrix Λ. Armed with an explicit approximation to $\mathrm{BR}(e, d_e^*)$, global multivariate optimisation over $e \in \mathcal{E}$ can be attempted; see the survey in [53].

Example 5.4 (Bayes A-, c- and E-optimality). Consider a loss $\ell(x, x') = \|x - x'\|_\Lambda^2$, where $\|x\|_\Lambda := \|\Lambda^{\frac{1}{2}}x\|_2$ and Λ is a positive semidefinite matrix with a square root $\Lambda^{\frac{1}{2}}$. Suppose that the posterior $\pi_{X|y,e}$ is a Gaussian $N(\mu_{y,e}, \Sigma_e)$. Then a Bayes decision rule d_e^* is defined through the Bayes act(s) a, which satisfy $\Lambda^{\frac{1}{2}}a = \Lambda^{\frac{1}{2}}\mu_{y,e}$ due to Proposition 5.2. Now observe that $\int \ell(x, a)\, d\pi_{X|y,e}(x) = \int (x - \mu_{y,e})^\top \Lambda (x - \mu_{y,e})\, d\pi_{X|y,e}(x) = \operatorname{tr}(\Lambda\Sigma_e)$, which is independent of y_e. It follows that $\mathrm{BR}(e, d_e^*) = \operatorname{tr}(\Lambda\Sigma_e)$. Selecting e to minimise $\operatorname{tr}(\Lambda\Sigma_e)$, or $\operatorname{tr}(\Sigma_e)$ in the common case, where $\Lambda = I$, is called *Bayes A-optimal* [7–10, 22, 47]. In the special case, where $\Lambda = cc^\top$ is a rank-1 matrix, the optimality criterion is called *Bayes c-optimality* [24]. Relatedly, minimisation of $\sup_{\|c\|=1} \operatorname{tr}(cc^\top\Sigma_e)$ is called *Bayes E-optimality* [11].

Example 5.5 (Bayes D-optimality). Consider a loss $\ell(x, x') = 0$ if $\|x - x'\|_\Lambda \le \epsilon$ and, otherwise, $\ell(x, x') = 1$. Suppose that the posterior $\pi_{X|y,e}$ is a Gaussian $N(\mu_{y,e}, \Sigma_e)$. A Bayes decision rule d_e^* is defined through the Bayes act(s), which one can verify include $\mu_{y,e}$. Now observe that $\int \ell(x, \mu_{y,e})\, d\pi_{X|y,e}(x) = \int 1_{\|x-\mu_{y,e}\|_\Lambda > \epsilon}\, d\pi_{X|y,e}(x)$, which is equal to one minus the probability that $\|Z\|_2 \le \epsilon$, where $Z \sim N(0, \Lambda^{1/2}\Sigma_e\Lambda^{1/2})$. For small ϵ, this is $1 - O(c_d^{-1} \det(\Lambda^{1/2}\Sigma_e\Lambda^{1/2})^{-d/2}\epsilon^d)$, where $c_d := 2^{\frac{d}{2}}\Gamma(\frac{d}{2} + 1)$. Note in particular that this is independent of y_e. It follows that $\mathrm{BR}(e, d_e^*)$ is minimised when $\det(\Lambda^{1/2}\Sigma_e\Lambda^{1/2})$ is minimised. This criterion is called *Bayes D-optimality* [56].

The notions above can be naturally extended to infinite-dimensional state spaces; see e. g. [1]. Given that the alphabet criteria are based on a Gaussian approximation of the posterior, it is incumbent on the analyst to attempt to verify the appropriateness of such approximations, for which a variety of diagnostics are available [15]. The alphabet criteria can also be derived as prior expectations of classical experimental design criteria, usually based on the Fisher information, an approach that was called *pseudo-Bayesian* in [53].

5.3.2 Design criteria via information theory

To perform Bayesian experimental design, a decision-theoretic basis is not essential. Indeed, other utilities—often based on information theory—have been proposed without reference to a decision-theoretic framework. For instance, [35] proposed

$$u(e, y) = D_{\mathrm{KL}}(\pi_{X|y,e} \| \pi_X),$$

where D_{KL} denotes the Kullback–Leibler divergence (or *information gain*) between the posterior and the prior.

This criterion is not explicitly motivated by estimation of a quantity of interest and may be better suited to tasks such as uncertainty quantification for the state $x \in \mathcal{X}$. However, it is interesting to note that the proposal of [35] can actually also be conceived from the decision-theoretic framework, albeit in a way that is nonstandard [5]. Indeed, consider an action set $\mathcal{A} = \mathcal{P}_{\mathcal{X}}$, whose elements $a \in \mathcal{A}$ are probability distributions on $\mathcal{X}$ and, for simplicity, existence of their Radon–Nikodym derivatives with respect to π_X is assumed. Consider also a loss function of the form

$$\ell(x, a) = -\log\left(\frac{\mathrm{d}a}{\mathrm{d}\pi_X}(x)\right) + f(x)$$

for arbitrary $f : \mathcal{X} \to \mathbb{R}$. Such a loss function is known as a *proper scoring rule*, since it can be shown that the unique Bayes act is to honestly report posterior belief; $a = \pi_{X|y,e}$ [5, Theorem 2]. Eq. (5.8) implies, in the particular case of $f \equiv 0$, that

$$\begin{aligned} u(e, y) &= -\int \ell(x, d_e^*(y))\, \mathrm{d}\pi_{X|y,e}(x) \\ &= \int \log\left(\frac{\mathrm{d}\pi_{X|y,e}}{\mathrm{d}\pi_X}(x)\right) \mathrm{d}\pi_{X|y,e}(x) \quad =: \; D_{\mathrm{KL}}(\pi_{X|y,e} \| \pi_X). \end{aligned}$$

The theory of proper scoring rules can be developed further, and the reader is referred to [23]. For other connections to information theory, we refer the reader to [26].

Next, we explicitly address the challenge of designing a utility function that is commensurate with the aims of a Bayesian probabilistic numerical method.

5.4 Optimality criteria for probabilistic numerical methods

The purpose of a Bayesian probabilistic numerical method is to provide formal uncertainty quantification for the state $x \in \mathcal{X}$ or a derived quantity of interest $\phi(x) \in \Phi$. To build intuition, we immediately provide an example of such distributional output:

Example 5.6 (Numerical integration, continued). Consider the numerical integration task, set out in Example 5.1. The output of a Bayesian probabilistic numerical method, with prior π_X equal to standard Weiner measure defined in Example 5.2 and information $y_e(x) = [x(t_0), \ldots, x(t_n)]$, is

$$d_e(y) = N\left(\sum_{i=1}^{n} \frac{(x(t_{i-1}) + x(t_i))}{2}(t_i - t_{i-1}), \sum_{i=1}^{n} \frac{(t_i - t_{i-1})^3}{12}\right).$$

An overview of Bayesian probabilistic numerical methods for integration is provided in [6].

The distributional output can be naturally reduced to a point estimator; for instance the mean is recognised as the trapezoidal rule, a classical numerical method. In this spirit of estimation, one can apply the alphabet criteria from Section 5.3.1; the univariate Gaussian nature of the output implies that A-, c-, E- and D- optimal experiments e all seek to minimise the variance $\frac{1}{12}\sum_{i=1}^{n}(t_i - t_{i-1})^3$. Thus, a decision-theoretic perspective leads to an optimal experiment with uniformly spaced nodes $t_i = \frac{i}{n}$.

However, the motivation for the development of probabilistic numerical methods is that the full distributional output provides formal uncertainty quantification for the quantity of interest. This objective is related *but not identical* to achieving strong performance at the estimation task. Indeed, for such uncertainty quantification to be useful, we put forward three desiderata to guide the design of a probabilistic numerical method d_e:

1. The output d_e should "put mass close" to the true quantity of interest ϕ, in a sense to be specified.
2. The output d_e should be "calibrated", in the sense that the true value $\phi(x)$ is well modelled as a random sample from the distributional output.
3. The search for an optimal experiment $e \in \mathcal{E}$ should, practically, not be too difficult.

Two important remarks are in order: First, notice the tension between (1) and (2), since (1) prefers an atomic mass placed on a "best" estimate, whereas (2) encourages probability density to be spread out so that the support at least contains the true value of the quantity of interest. Second, observe that for a *Bayesian* probabilistic numerical method it is not possible in general to ensure (2) through suitable choice of experiment $e \in \mathcal{E}$, since the output d_e depends crucially on the prior π_X. Hence, our focus in the remainder is on satisfying (1) and (3) for a Bayesian probabilistic numerical method, with (2) being assumed through suitable elicitation of π_X. A detailed discussion of prior elicitation was provided in the supplement of [18] and sophisticated approaches to data-dependent elicitation have recently been developed, e. g. [30].

Remark 5.2 (Dependence on π_X). The dependence of a Bayesian decision rule d_e on the prior π_X is a fundamental aspect of Bayesian decision theory, thus the difficulties just described with regard to (2) are also fundamental. The alternative approach to decision theory pursued by Wald [60] circumvents the elicitation of a prior π_X by casting the decision problem in an adversarial context, such that an adversary selects $\tilde{\pi}_X$, which need not coincide with π_X, from which $X \sim \tilde{\pi}_X$ is generated. The decision-maker then adopts a Bayes rule d_e^*, as in (5.4), based on a prior π_X that solves the minimax optimisation problem

$$\min_{\pi_X} \max_{\tilde{\pi}_X} \iint \ell(x, d_e^*(y)) \, \mathrm{d}\pi_{Y|x,e}(y) \, \mathrm{d}\tilde{\pi}_X(x).$$

Wald's approach is considered in detail in the context of probabilistic numerical methods in [48, 49], and forms an interesting point of contrast to the present discussion.

5.4.1 The BPN criterion

In what follows, we consider the action space $\mathcal{A}$ to be equal to the state space $\mathcal{X}$. In that context, [18] proposed the utility

$$u(e, x, y) = -\int \ell(x, x') \, \mathrm{d}\pi_{X|y,e}(x'), \tag{5.9}$$

which differs from eq. (5.8) in that it depends explicitly on the true state x and measures the posterior expected loss with respect to x. Thus, two posteriors that give rise to the same Bayes act(s) need not give rise to the same utility function in eq. (5.9); see Figure 5.2. The associated experimental design criterion, denoted BPN, is a special case of eq. (5.7) with the above utility:

$$\mathrm{BPN}(e) := \iiint \ell(x, x') \, \mathrm{d}\pi_{X|y,e}(x') \, \mathrm{d}\pi_{Y|x,e}(y) \, \mathrm{d}\pi_X(x). \tag{5.10}$$

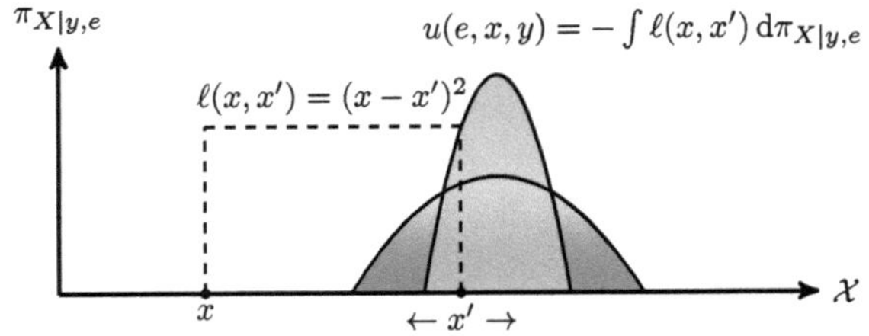

Figure 5.2: The BPN criterion is motivated by the idea that a Bayesian probabilistic numerical method should concentrate its mass close to the true state (x). In that respect, the two posteriors ($\pi_{X|y,e}$; blue and green densities) lead to different utilities (u) in general. In this illustration, the more concentrated (green) posterior is preferred by the BPN criterion; in particular, the BPN criterion makes no attempt to ensure that the posterior is well calibrated. (The calibration of the posterior depends chiefly on whether or not an appropriate prior π_X is elicited.)

To build intuition, for $\mathcal{X}$, a metric space with metric d and $\ell(x,x') = d(x,x')^p$, the inner integral

$$\int \ell(x,x')\,\mathrm{d}\pi_{X|y,e}(x') = D_{\mathrm{W},p}(\delta(x),\pi_{X|y,e}) \tag{5.11}$$

is the pth Wasserstein distance between the posterior $\pi_{X|y,e}$ and an atom on the true state $\delta(x)$. Thus, it is seen that desideratum (1) is satisfied. Note also that desideratum (3) is satisfied. Indeed, in contrast to the decision-theoretic approach to experimental design, where Gaussian approximations, such as the alphabet criteria, are practically necessary for an optimal experiment to be computed (to circumvent the repeated calculation of a Bayes act), no Gaussian approximation is needed to numerically compute $\mathrm{BPN}(e)$. Indeed, we can rewrite the BPN criterion in eq. (5.10) as

$$\mathrm{BPN}(e) = \int \underbrace{\iint \ell(x,x')\,\mathrm{d}\pi_{X|y,e}(x')\,\mathrm{d}\pi_{X|y,e}(x)}_{(*)}\,\mathrm{d}\pi_{Y|e}(y). \tag{5.12}$$

From this perspective it is clear that $\mathrm{BPN}(e)$ can be directly computed using a Monte Carlo method to sample from $\pi_{X|y,e}$ in order to approximate the inner integral $(*)$. Although not a closed form, we argue that the sophistication of modern Monte Carlo methods mean that desideratum (3) is essentially satisfied. The intergal $(*)$ has been termed a *distance-based information function* in [26] for the case $\ell(x,x') = \|\phi(x) - \phi(x')\|_\Phi^2$.

Example 5.7 (Numerical integration, continued). Unpacking Example 5.6, we note that $\pi_{X|y,e}$ is a collection of independent Brownian bridges $X_i\colon [t_{i-1},t_i] \to \mathbb{R}$ with $X_i(t_{i-1}) = x(t_{i-1})$, $X_i(t_i) = x(t_i)$ and covariance function $\mathrm{cov}(X_i(t),X_i(t')) = \frac{(t_i-t')(t-t_{i-1})}{(t_i-t_{i-1})}$, $t_{i-1} \le t \le t' \le t_i$. Through direct calculation, we obtain the closed form $\mathrm{BPN}(e) = \frac{1}{6}\sum_{i=1}^n (t_i - t_{i-1})^3$, which is easily minimised to obtain an optimal experiment by taking $t_i = \frac{i}{n}$. That this should coincide with the BDT-optimal experiment is explained later in Proposition 5.3.

In general, the set of optimal experiments according to this criterion will be denoted

$$\mathcal{E}^*_{\mathrm{BPN}} := \operatorname*{arg\,min}_{e\in\mathcal{E}} \mathrm{BPN}(e).$$

For the remainder, our focus is on the properties of the BPN criterion, and in particular we will ask, for this choice, whether $\mathcal{E}^*_{\mathrm{BPN}} \stackrel{?}{=} \mathcal{E}^*_{\mathrm{BDT}}$.

5.4.2 Numerical illustration

It was explained in Section 5.4.1 that the BPN criterion can be computed in circumstances that are quite general, as opposed to the BDT criteria, for which Gaussian approximations are typically needed. The aim of this section is to illustrate the use of the BPN criteria on a problem, for which BDT cannot easily be applied.

Consider the canonical linear elliptic partial differential equation

$$\begin{aligned} -\Delta x(t) &= f(t), \quad t \in D \\ x(t) &= g(t), \quad t \in \partial D \end{aligned}$$

on $D := (0,1)^2$. Let $\ell(x,x') = (\int (x(t) - x'(t))^p \, \mathrm{d}t)^{1/p}$, and let the prior π_X be Gaussian with $\int x(t) \, \mathrm{d}\pi_X(x) = 0$, $\int x(t)x(t') \, \mathrm{d}\pi_X(x) = \exp(-\|t - t'\|_2^2)$. An experiment $e \in \mathcal{E}$ determines the set of locations $\{t_i\}_{i=1}^m \subset D$, $\{t_i\}_{i=m+1}^n \subset \partial D$, at which the functions f and g, respectively, are evaluated. To simplify the presentation, we suppose that $m \ll n$ and focus on the information bottleneck, which is the placement of the points $\{t_i\}_{i=1}^m$ in D. To minimise BPN(e), we recognised the triple integral in eq. (5.12) as a single joint Gaussian integral and, to approximate this integral, we employed the standard Monte Carlo method.

For $p = 2$, it can be shown that the optimal experiment, according to BPN, coincides with the optimal experiment, according to BDT (see Proposition 5.3). The points $\{t_i\}_{i=1}^m$ were optimised sequentially in Figure 5.3 for convenience, and we note that a space-filling design was obtained. For $p = \infty$, in contrast, it is clear from Figure 5.4 that the optimal design from BPN is no longer space-filling. However, since it is not easily computed, it is unclear whether or not the optimal experiment, according to BPN, resembles the optimal experiment according to BDT.

5.4.3 Analysis and open questions

Our final aim is to address the question of whether $\mathcal{E}^*_{\mathrm{BPN}} \stackrel{?}{=} \mathcal{E}^*_{\mathrm{BDT}}$. To this end, we first present a positive result in Proposition 5.3, and then present a negative result in Proposition 5.4.

5.4.3.1 A positive result

Under an assumption on the form of the loss ℓ, the following equivalence was established in [18]:

Proposition 5.3. *Consider a loss function of the form $\ell(x,x') = \|\phi(x) - \phi(x')\|_\Phi^2$, where $\phi: \mathcal{X} \to \Phi$ takes values in an inner product space Φ, with inner product $\langle \cdot, \cdot \rangle_\Phi$ and induced norm $\|\varphi\|_\Phi = \langle \varphi, \varphi \rangle_\Phi^{1/2}$. Suppose that the first conclusion of Proposition 5.2 holds (i. e. eq. (5.6)). Then $\mathcal{E}^*_{\mathrm{BPN}} = \mathcal{E}^*_{\mathrm{BDT}}$.*

For completeness, a slightly more concise proof of this result is included in Appendix A.3. This form of loss is commonly encountered and includes the Wasserstein

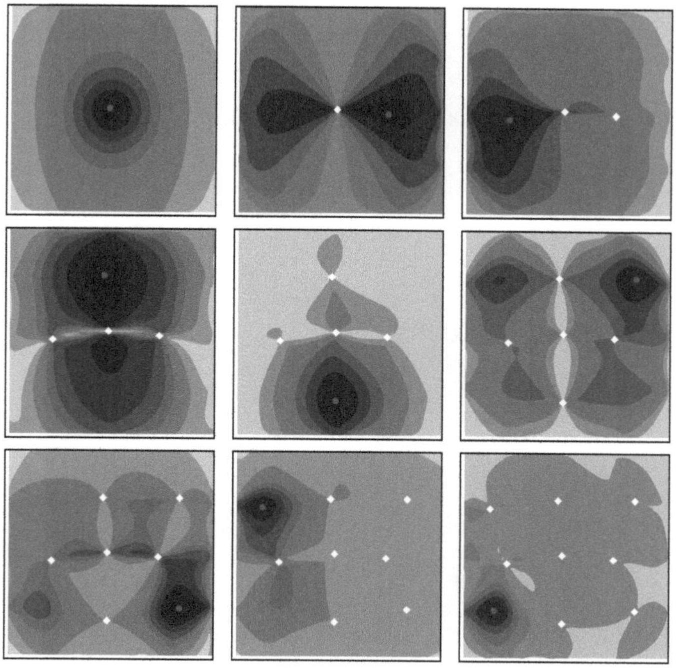

Figure 5.3: Optimal experimental design with the BPN criterion, with $p = 2$ used. (The filled contours depict BPN(e) as a function of t_m where the lexicographic ordering of the panels corresponds to $m = 1, \ldots, 9$. The white points represent the fixed values of $\{t_i\}_{i=1}^{m-1}$ and the red points represent the minimal value of t_m.)

distance with $p = 2$ as a particular instance (see eq. (5.11)). A consequence of Proposition 5.3 is that one can minimise the BPN criterion as an alternative to the BDT criteria in circumstances where the hypotheses of the Proposition hold.

5.4.3.2 A negative result

A converse result to Proposition 5.3 can also be constructed. The following is inspired by, but slightly more elegant than, a corresponding result in [18]:

Proposition 5.4. *Suppose that the state space $\mathcal{X}$ can be partitioned into three disjoint subsets, each with positive probability under π_X. Then there exists a loss function ℓ and a set of candidate experiments $\mathcal{E}$ such that $\mathcal{E}^*_{\mathrm{BPN}} \neq \mathcal{E}^*_{\mathrm{BDT}}$.*

The proof of this result is constructive and is included in Appendix A.4. It is clear that the assumptions on $\mathcal{X}$ and π_X are weak, to the point of being trivial.

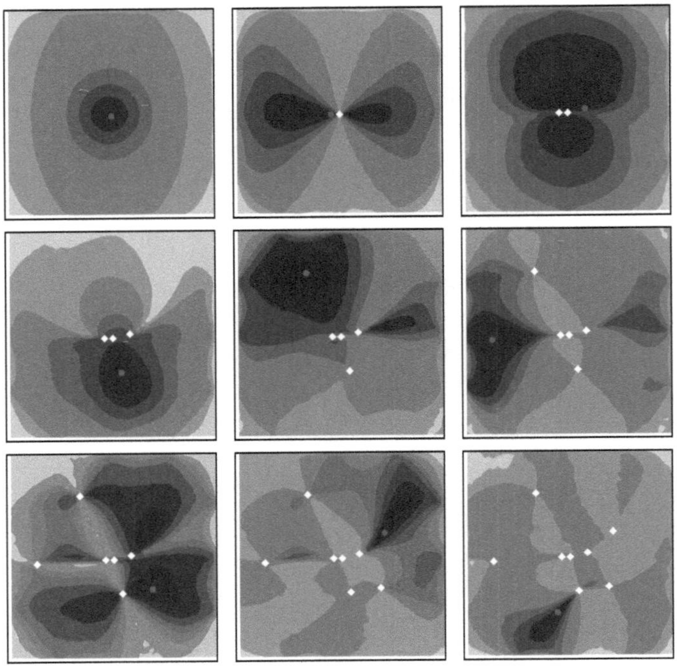

Figure 5.4: Optimal experimental design with the BPN criterion, with $p = \infty$ used. (The filled contours depict BPN(e) as a function of t_m where the lexicographic ordering of the panels corresponds to $m = 1, \ldots, 9$. The white points represent the fixed values of $\{t_i\}_{i=1}^{m-1}$ and the red points represent the minimal value of t_m.)

5.4.3.3 Open questions

An explicit characterization of the loss functions ℓ, for which $\mathcal{E}^*_{\mathrm{BPN}} = \mathcal{E}^*_{\mathrm{BDT}}$ is not, at least to our knowledge, available at present. In particular, the analytic intractability of optimal experiments in all but the simplest of numerical tasks leaves it unclear whether there exist a numerical task of practical importance for which $\mathcal{E}^*_{\mathrm{BPN}} \neq \mathcal{E}^*_{\mathrm{BDT}}$.

In general, any utility function u from the Bayesian experimental design literature provides a criterion BED(e) that can be studied from an information-based complexity standpoint. In particular, the issue of how BED(e^*) scales with $\dim(\mathcal{X})$ when $e^* \in \mathcal{E}^*_{\mathrm{BED}}$, so-called *tractability*, can be studied.

5.5 Discussion

The aim of this article was to build on the earlier work of [31], drawing attention to a wider range of optimality criteria that are used in the Bayesian experimental design literature, and considering their use in the numerical context. In particular, we explained how the competing objectives of parameter estimation and uncertainty quantification lead, in general, to different notions of optimal information for a probabilistic numerical method.

One criterion, called BPN, was explored in detail. This formalised the idea that probability mass ought to be located close to the true quantity of interest. However, several other factors also relevant in the design of a probabilistic numerical method were not discussed. Indeed, the BPN criterion does not encode the notion that a posterior ought to be well calibrated, nor the notion that inferences should be robust to prior misspecification, nor does it attempt more nuanced control (e. g., at the wall clock level) of computational cost [59]. In addition, no attempt was made to extend the notions introduced in this work to the adaptive context, where notions from sequential Bayesian experimental design are needed [25]. These are active areas of research, and we look forward to seeing how these ideas are developed.

Appendix A. Proofs

This appendix collects together proofs for the results quoted in the main text.

A.1 Proof of Proposition 5.1

Let $d_e^\circ \in \mathcal{D}_e$ be any decision rule such that $d_e^\circ(y) \in \mathcal{A}_e^*(y)$ for all $y \in \mathcal{Y}$. Then, for any other $d_e \in \mathcal{D}_e$,

$$\begin{aligned}
\mathrm{BR}(e, d_e^\circ) - \mathrm{BR}(e, d_e) &= \iint \ell(x, d_e^\circ(y)) - \ell(x, d_e(y))\, \mathrm{d}\pi_{Y|x,e}(y)\, \mathrm{d}\pi_X(x) \\
&= \int \Big(\int \ell(x, d_e^\circ(y)) - \ell(x, d_e(y))\, \mathrm{d}\pi_{X|y,e}(x) \Big)\, \mathrm{d}\pi_{Y|e}(y) \\
&\le \int 0\, \mathrm{d}\pi_{Y|e}(y) \; = \; 0.
\end{aligned}$$

Thus, $\mathrm{BR}(e, d_e^\circ) \le \mathrm{BR}(e, d_e)$ and d_e° is a Bayes rule. Conversely, let $d_e^* \in \mathcal{D}_e$ be a Bayes rule. Then

$$\begin{aligned}
0 &\ge \iint \ell(x, d_e^*(y)) - \ell(x, d_e^\circ(y))\, \mathrm{d}\pi_{Y|x,e}(y)\, \mathrm{d}\pi_X(x) \\
&= \int \underbrace{\Big(\int \ell(x, d_e^*(y)) - \ell(x, d_e^\circ(y))\, \mathrm{d}\pi_{X|y,e}(x) \Big)}_{(*)}\, \mathrm{d}\pi_{Y|e}(y).
\end{aligned}$$

The definition of d_e° implies that $(*) \ge 0$. Thus, $\ell(x, d_e^*(y)) = \ell(x, d_e^\circ(y))$ holds $\pi_{Y|e}$-almost everywhere, and so $d_e^*(y) \in \mathcal{A}_e^*(y)$ for $\pi_{Y|e}$-almost all $y \in \mathcal{Y}_e$, as required.

A.2 Proof of Proposition 5.2

Fix $e \in \mathcal{E}$ and $y_e \in \mathcal{Y}_e$. Consider the function

$$f(a) := \int \ell(x,a)\,\mathrm{d}\pi_{X|y,e}(x)$$

and recall that the set $\mathcal{A}_e^*(y_e)$ of Bayes acts is defined as the elements $a \in \mathcal{A}$, for which $f(a)$ is minimised. The assumptions on ℓ and ϕ imply that f can be twice differentiated. The first derivative is easily computed, the integrability assumption on ϕ permits differentiation under the integral sign, yielding

$$\begin{aligned}\frac{\mathrm{d}f}{\mathrm{d}a} &= \int \frac{\mathrm{d}}{\mathrm{d}a}\ell(x,a)\,\mathrm{d}\pi_{X|y,e}(x)\\ &= \int\left\{-2\frac{\mathrm{d}\phi}{\mathrm{d}a}(a)^\top[\phi(x)-\phi(a)]\right\}\mathrm{d}\pi_{X|y,e}(x)\\ &= -2\frac{\mathrm{d}\phi}{\mathrm{d}a}(a)^\top \underbrace{\int[\phi(x)-\phi(a)]\,\mathrm{d}\pi_{X|y,e}(x)}_{(*)}.\end{aligned}$$

Since the matrix $\frac{\mathrm{d}\phi}{\mathrm{d}a}$ has full row rank, and since the space $\mathcal{A}$ is without boundary, if a locally extremises f, then the term $(*)$ must vanish. This last condition is precisely eq. (5.6).

For the converse result, assume that $\pi_{X|y,e}$ is a nonatomic distribution on $\mathcal{X}$, since otherwise the result is trivial. Let $\frac{\partial^2\phi}{\partial a_i \partial a_j}$ be the m-dimensional column vector with entries $\frac{\partial^2\phi_k}{\partial a_i\partial a_j}$. Then f is seen to have continuous second derivative

$$\begin{aligned}\left[\frac{\mathrm{d}^2 f}{\mathrm{d}a^2}\right]_{i,j} &= \frac{\partial^2 f}{\partial a_i\partial a_j}\\ &= -2\sum_{k=1}^m \frac{\partial^2\phi_k}{\partial a_i\partial a_j}\int[\phi_k(x)-\phi_k(a)]\,\mathrm{d}\pi_{X|y,e}(x) + 2\sum_{k=1}^m \frac{\partial\phi_k}{\partial a_i}\frac{\partial\phi_k}{\partial a_j}\\ &= -2\frac{\partial^2\phi}{\partial a_i\partial a_j}^\top\int[\phi(x)-\phi(a)]\,\mathrm{d}\pi_{X|y,e}(x) + 2\left[\frac{\mathrm{d}\phi}{\mathrm{d}a}^\top\frac{\mathrm{d}\phi}{\mathrm{d}a}\right]_{i,j},\end{aligned}$$

since ϕ is assumed to be twice continuously differentiable. The result will follow if f is coercive, since a unique local extremum of a coercive function with continuous second derivatives must be a global minimum of that function. To this end, consider the reverse triangle inequality

$$\begin{aligned}f(a) &= \int\|\phi(x)-\phi(a)\|_2^2\,\mathrm{d}\pi_{X|y,e}(x)\\ &\geq \int\big|\|\phi(x)\|_2 - \|\phi(a)\|_2\big|^2\,\mathrm{d}\pi_{X|y,e}(x)\end{aligned}$$

$$= \underbrace{\int \|\phi(x)\|_2^2 \, \mathrm{d}\pi_{X|y,e}(x)}_{c_1} - 2\|\phi(a)\|_2 \underbrace{\int \|\phi(x)\|_2 \, \mathrm{d}\pi_{X|y,e}(x)}_{c_2} + \|\phi(a)\|_2^2$$
$$=: g(\|\phi(a)\|_2),$$

where g is the quadratic $g(z) = c_1 - 2c_2 z + z^2$. From Jensen's inequality, $c_2^2 \leq c_1$ with strict inequality if $\pi_{X|y,e}$ is not an atomic distribution on $\mathcal{X}$. Thus, the coercivity of ϕ implies the coercivity of f, and the proof is complete.

A.3 Proof of Proposition 5.3

To start, we indicate how the assumptions on the loss function will be used. Observe that

$$\iint \|\phi(x) - \phi(x')\|_\Phi^2 \, \mathrm{d}\pi_{X|y,e}(x) \, \mathrm{d}\pi_{X|y,e}(x') = 2 \int \|\phi(x) - \bar{\phi}_{y,e}\|_\Phi^2 \, \mathrm{d}\pi_{X|y,e}(x), \tag{5.13}$$

where $\bar{\phi}_{y,e} = \int \phi(x) \, \mathrm{d}\pi_{X|y,e}(x)$. Indeed,

$$\phi(x) - \phi(x') = (\phi(x) - \bar{\phi}_{y,e}) - (\phi(x') - \bar{\phi}_{y,e}),$$

and

$$\|\phi(x) - \phi(x')\|_\Phi^2 = \|\phi(x) - \bar{\phi}_{y,e}\|_\Phi^2 - 2\langle \phi(x) - \bar{\phi}_{y,e}, \phi(x') - \bar{\phi}_{y,e} \rangle_\Phi + \|\phi(x') - \bar{\phi}_{y,e}\|_\Phi^2,$$

where the first and third terms have identical integrals under $\mathrm{d}\pi_{X|y,e}(x) \, \mathrm{d}\pi_{X|y,e}(x')$ and

$$\int \langle \phi(x) - \bar{\phi}_{y,e}, \phi(x') - \bar{\phi}_{y,e} \rangle_\Phi \, \mathrm{d}\pi_{X|y,e}(x) = \langle 0, \phi(x') - \bar{\phi}_{y,e} \rangle_\Phi = 0.$$

Next, we use the fact that $\mathrm{d}\pi_{Y|x,e}(y) \, \mathrm{d}\pi_X(x) = \mathrm{d}\pi_{X|y,e}(x) \, \mathrm{d}\pi_{Y|e}(y)$ to observe that, for the loss $\ell(x, x') = \|\phi(x) - \phi(x')\|_\Phi^2$,

$$\begin{aligned}
&\mathrm{BPN}(e) \\
&= \iiint \|\phi(x) - \phi(x')\|_\Phi^2 \, \mathrm{d}\pi_{X|y,e}(x') \, \mathrm{d}\pi_{Y|x,e}(y) \, \mathrm{d}\pi_X(x) \quad \text{(from Eq. (5.10))} \\
&= \int \left(\iint \|\phi(x) - \phi(x')\|_\Phi^2 \, \mathrm{d}\pi_{X|y,e}(x) \, \mathrm{d}\pi_{X|y,e}(x') \right) \mathrm{d}\pi_{Y|e}(y) && (5.14) \\
&= \int \left(2 \int \|\phi(x) - \bar{\phi}_{y,e}\|_\Phi^2 \, \mathrm{d}\pi_{X|y,e}(x) \right) \mathrm{d}\pi_{Y|e}(y) \quad \text{(from eq. (5.13))}. && (5.15)
\end{aligned}$$

Third, we recall that, for the loss $\ell(x, x') = \|\phi(x) - \phi(x')\|_\Phi^2$, a Bayes act $a \in \mathcal{A}$ must satisfy $\phi(a) = \bar{\phi}_{y,e}$ under the first conclusion of Proposition 5.2 (i. e., eq. (5.6)). Therefore,

$$\begin{aligned}
\mathrm{BR}(e, d_e^*) &= \iint \|\phi(x) - \bar{\phi}_{y,e}\|_\Phi^2 \, \mathrm{d}\pi_{Y|x,e}(y) \, \mathrm{d}\pi_X(x) \quad \text{(from eq. (5.3))} \\
&= \iint \|\phi(x) - \bar{\phi}_{y,e}\|_\Phi^2 \, \mathrm{d}\pi_{X|y,e}(x) \, \mathrm{d}\pi_{Y|e}(y), && (5.16)
\end{aligned}$$

where the final equality follows the same argument used to obtain eq. (5.14). This final expression, in eq. (5.16), is observed to be exactly half of the expression for BPN(e) obtained in eq. (5.15). Therefore, minimisation of eq. (5.16) is equivalent to minimisation of eq. (5.15), and it follows that $\mathcal{E}^*_{\text{BDT}} = \mathcal{E}^*_{\text{BPN}}$, as claimed.

A.4 Proof of Proposition 5.4

To start, suppose we are presented with a state space $\mathcal{X}$, that can be partitioned into three measurable subsets ①, ②, and ③, with respective strictly positive probabilities $\pi_{①}$, $\pi_{②}$, and $\pi_{③}$. Without loss of generality, we assume that $0 < \pi_{①} \le \pi_{②} \le \pi_{③} < 1$, noting that $\pi_{①} + \pi_{②} + \pi_{③} = 1$. In what follows, we exhibit a collection of candidate experiments $\mathcal{E}$ that, in effect, reveal some information about the state x at the coarse level of the partition {①, ②, ③}. Let $s: \mathcal{X} \to \{①, ②, ③\}$ be the map that assigns each $x \in \mathcal{X}$ to its partition element. Consider a set of just two candidate experiments, in each case reporting a deterministic observation of the latent state:

$$\mathcal{E} = \begin{Bmatrix} e_1 \ : \ y(x) = 1_{s(x)\in\{①\}} \\ e_2 \ : \ y(x) = 1_{s(x)\in\{①,②\}} \end{Bmatrix}.$$

The aim is to decide whether experiment e_1 or e_2 should be performed. The informativeness of an experiment will be measured as described in the main text, based on the so-called 0–1 loss

$$\ell(x, x') = \begin{cases} 0 & \text{if } 1_{s(x)=①} = 1_{s(x')=①}, \\ 1 & \text{otherwise.} \end{cases}$$

The 0–1 loss depends on the state x only through the indicator function $1_{s(x)=①}$, so that our task is equivalent to guessing whether the true $s(x)$ is ① or not, based on information obtained in the experiment. Thus, in a small abuse of notation, we may redefine the action space to be $\mathcal{A} = \{①, \neg①\}$.

A.4.1 Bayesian decision theory

First we derive the BDT-optimal experiment(s) $\mathcal{E}^*_{\text{BDT}}$. Without loss of generality, we can restrict attention, in the search for a Bayes rule, to the set of nonrandom decision rules of the form $d_e(y) = 1_{y=0}a_0 + 1_{y=1}a_1$ for some $a_1, a_2 \in \mathcal{A}$ to be specified. The actions a_0, a_1 are selected to minimise the risk

$$\text{BR}(e, d_e) = \int [y(x)\ell(x, a_1) + (1 - y(x))\ell(x, a_0)] \, \mathrm{d}\pi_X(x).$$

Inputting the form of each experiment, we obtain

$$\text{BR}(e_1, d_{e_1}) = \pi_{①} 1_{a_1 \neq ①} + \pi_{②} 1_{a_0 = ①} + \pi_{③} 1_{a_0 = ①}$$

$$\mathrm{BR}(e_2, d_{e_2}) = \pi_{①} 1_{a_1 \neq ①} + \pi_{②} 1_{a_1 = ①} + \pi_{③} 1_{a_0 = ①}.$$

The expression for $\mathrm{BR}(e_1, d_{e_1})$ is minimised by $a_1 = \neg ①$ and $a_0 = \neg ①$, since $\pi_{①}, \pi_{②}, \pi_{③} > 0$. Similarly, the expression for $\mathrm{BR}(e_2, d_{e_2})$ is minimised by $a_1 = \neg ①$ and $a_0 = \neg ①$, since $\pi_{①} \le \pi_{②}$ and $\pi_{③} > 0$. It follows that $\mathrm{BR}(e_2, d^*_{e_2}) = \mathrm{BR}(e_1, d^*_{e_1}) = \pi_{①}$, and both experiments are optimal. Therefore, $\mathcal{E}^*_{\mathrm{BDT}} = \{e_1, e_2\}$.

A.4.2 Bayesian experimental design

Now we derive the BPN-optimal experiment(s) $\mathcal{E}^*_{\mathrm{BPN}}$. Through direct computation, we see that

$$\begin{aligned}
\mathrm{BPN}(e_1) &= \pi_{①} \int 1_{s(x') \neq ①} \, \mathrm{d}\pi_{X|y=1,e_1}(x') + \pi_{②} \int 1_{s(x') = ①} \, \mathrm{d}\pi_{X|y=0,e_1}(x') \\
&\quad + \pi_{③} \int 1_{s(x') = ①} \, \mathrm{d}\pi_{X|y=0,e_1}(x') \\
&= \pi_{①} \times 0 + \pi_{②} \times 0 + \pi_{③} \times 0 = 0 \\
\mathrm{BPN}(e_2) &= \pi_{①} \int 1_{s(x') \neq ①} \, \mathrm{d}\pi_{X|y=1,e_2}(x') + \pi_{②} \int 1_{s(x') = ①} \, \mathrm{d}\pi_{X|y=1,e_2}(x') \\
&\quad + \pi_{③} \int 1_{s(x') = ①} \, \mathrm{d}\pi_{X|y=0,e_2}(x') \\
&= \pi_{①} \frac{\pi_{②}}{\pi_{①} + \pi_{②}} + \pi_{②} \frac{\pi_{①}}{\pi_{①} + \pi_{②}} + \pi_{③} \times 0 = \frac{2\pi_{①}\pi_{②}}{\pi_{①} + \pi_{②}}.
\end{aligned}$$

Since $\pi_{②} \ge \pi_{①} > 0$, it follows that experiment e_1 is the *only* optimal experiment. Therefore, $\mathcal{E}^*_{\mathrm{BPN}} = \{e_1\} \neq \{e_1, e_2\} = \mathcal{E}^*_{\mathrm{BDT}}$, as claimed.

Bibliography

[1] A. Alexanderian, P. J. Gloor, and O. Ghattas. On Bayesian A- and D-optimal experimental designs in infinite dimensions. *Bayesian Analysis*, 11(3):671–695, 2016.

[2] B. Amzal, F. Y. Bois, E. Parent, and C. P. Robert. Bayesian-optimal design via interacting particle systems. *Journal of the American Statistical Association*, 101(474):773–785, 2006.

[3] S. Bartels, J. Cockayne, I. C. F. Ipsen, and P. Hennig. Probabilistic linear solvers: A unifying view. *Statistics and Computing*, 2019. To appear.

[4] J. O. Berger. *Statistical decision theory and Bayesian analysis*. Springer Science & Business Media, 1985.

[5] J. M. Bernardo. Expected information as expected utility. *Annals of Statistics*, 7(3):686–690, 1979.

[6] F.-X. Briol, C. J. Oates, M. Girolami, M. A. Osborne, and D. Sejdinovic. Probabilistic integration: A role in statistical computation? (with discussion and rejoinder). *Statistical Science*, 34(1):1–22, 2019. Discussion and rejoinder on p. 23–42.

[7] R. J. Brooks. A decision theory approach to optimal regression designs. *Biometrika*, 59(3):563–571, 1972.

[8] R. J. Brooks. On the choice of an experiment for prediction in linear regression. *Biometrika*, 61(2):303–311, 1974.
[9] R. J. Brooks. Optimal regression designs for prediction when prior knowledge is available. *Metrika*, 23(1):221–230, 1976.
[10] R. J. Brooks. Optimal regression design for control in linear regression. *Biometrika*, 64(2):319–325, 1977.
[11] K. Chaloner. Optimal Bayesian experimental design for linear models. *Annals of Statistics*, pages 283–300, 1984.
[12] K. Chaloner and I. Verdinelli. Bayesian experimental design: A review. *Statistical Science*, pages 273–304, 1995.
[13] J. T. Chang and D. Pollard. Conditioning as disintegration. *Statistica Neerlandica*, 51(3):287–317, 1997.
[14] Y. Chen, A. Huang, Z. Wang, I. Antonoglou, J. Schrittwieser, D. Silver, and N. de Freitas. Bayesian optimization in AlphaGo. *arXiv:1812.06855*, 2018.
[15] M. A. Clyde. *Bayesian optimal designs for approximate normality*. PhD thesis, University of Minnesota, 1993.
[16] M. A. Clyde. Experimental design: A Bayesian perspective. *International Encyclopia Social and Behavioral Sciences*, 2001.
[17] J. Cockayne, C. Oates, I. Ipsen, and M. Girolami. A Bayesian conjugate gradient method (with discussion and rejoinder). *Bayesian Analysis*, 2019. To appear.
[18] J. Cockayne, C. Oates, T. J. Sullivan, and M. Girolami. Bayesian probabilistic numerical methods. *SIAM Review*, 2019. To appear.
[19] J. Cockayne, C. J. Oates, T. J. Sullivan, and M. Girolami. Probabilistic numerical methods for PDE-constrained Bayesian inverse problems. *arXiv:1605.07811*, 2017.
[20] A. P. Dawid and P. Sebastiani. Coherent dispersion criteria for optimal experimental design. *Annals of Statistics*, 27(1):65–81, 1999.
[21] P. Diaconis. Bayesian numerical analysis. In *Statistical Decision Theory and Related Topics IV, volume 1*, pages 163–175. Springer-Verlag New York, 1988.
[22] G. Duncan and M. H. DeGroot. A mean squared error approach to optimal design theory. In *Proceedings of the 1976 Conference on Information: Science and Systems*, pages 217–221, 1976.
[23] W. Ehm and T. Gneiting. Local proper scoring rules of order two. *Annals of Statistics*, 40(1):609–637, 2012.
[24] S. M. El-Krunz and W. J. Studden. Bayesian optimal designs for linear regression models. *Annals of Statistics*, 19(4):2183–2208, 1991.
[25] B. K. Ghosh and P. K. Sen. *Handbook of Sequential Analysis*. CRC Press, 1991.
[26] M. Hainy, W. G. Müller, and H. Wynn. Learning functions and approximate Bayesian computation design: ABCD. *Entropy*, 16(8):4353–4374, 2014.
[27] P. Hennig. Probabilistic interpretation of linear solvers. *SIAM Journal on Optimization*, 25(1):234–260, 2015.
[28] P. Hennig and M. Kiefel. Quasi-Newton methods: A new direction. *Journal of Machine Learning Research*, 14:843–865, 2013.
[29] P. Hennig, M. A. Osborne, and M. Girolami. Probabilistic numerics and uncertainty in computations. *Proceedings of the Royal Society A*, 471(2179):20150142, 2015.
[30] R. Jagadeeswaran and F. J. Hickernell. Fast automatic Bayesian cubature using lattice sampling. *arXiv:1809.09803*, 2018.
[31] J. B. Kadane and G. W. Wasilkowski. Average Case ϵ-Complexity in Computer Science: A Bayesian View. In *Bayesian Statistics*, pages 361–374. Elsevier, North-Holland, 1985.

[32] T. Karvonen, C. J. Oates, and S. Särkkä. A Bayes–Sard cubature method. In *Proceedings of the 32nd Conference on Neural Information Processing Systems*, 2018.
[33] F. M. Larkin. Gaussian measure in Hilbert space and applications in numerical analysis. *The Rocky Mountain Journal of Mathematics*, 2(3):379–421, 1972.
[34] F. M. Larkin. Probabilistic error estimates in spline interpolation and quadrature. In *Information Processing 74 (Proceedings of IFIP Congress, Stockholm, 1974), volume 74*, pages 605–609. North-Holland, 1974.
[35] D. V. Lindley. On a measure of the information provided by an experiment. *Annals of Mathematical Statistics*, 27(4):986–1005, 1956.
[36] D. V. Lindley. *Bayesian Statistics, A Review*. Society for Industrial and Applied Mathematics, 1972.
[37] M. Mahsereci and P. Hennig. Probabilistic line searches for stochastic optimization. In *Proceedings of the 29th Conference on Neural Information Processing Systems*, 2015.
[38] J. Mockus. *Bayesian Approach to Global Optimization: Theory and Applications*. Springer Science & Business Media, 1989.
[39] P. Müller. Simulation based optimal design. *Handbook of Statistics*, 25:509–518, 2005.
[40] C. J. Oates, J. Cockayne, R. G. Aykroyd, and M. Girolami. Bayesian probabilistic numerical methods in time-dependent state estimation for industrial hydrocyclone equipment. *Journal of the American Statistical Association*, 2019. To appear.
[41] J. Oettershagen. *Construction of Optimal Cubature Algorithms with Applications to Econometrics and Uncertainty Quantification*. PhD thesis, Institut für Numerische Simulation, Universität Bonn, 2017.
[42] A. O'Hagan. Curve fitting and optimal design for prediction (with discussion). *Journal of the Royal Statistical Society, Series B*, 40(1):1–42, 1978.
[43] A. O'Hagan. Bayes–Hermite quadrature. *Journal of Statistical Planning and Inference*, 29(3):245–260, 1991.
[44] A. O'Hagan. Some Bayesian numerical analysis. *Bayesian Statistics*, 4:345–363, 1992.
[45] A. M. Overstall, J. M. McGree, and C. C. Drovandi. An approach for finding fully Bayesian optimal designs using normal-based approximations to loss functions. *Statistics and Computing*, 28(2):343–358, 2018.
[46] A. M. Overstall and D. C. Woods. Bayesian design of experiments using approximate coordinate exchange. *Technometrics*, 59:458–470, 2017.
[47] R. J. Owen. The optimum design of a two-factor experiment using prior information. *Annals of Mathematical Statistics*, 41(6):1917–1934, 1970.
[48] H. Owhadi. Bayesian numerical homogenization. *Multiscale Modeling & Simulation*, 13(3):812–828, 2015.
[49] H. Owhadi. Multigrid with rough coefficients and multiresolution operator decomposition from hierarchical information games. *SIAM Review*, 59(1):99–149, 2017.
[50] J. Prüher, T. Karvonen, C. J. Oates, O. Straka, and S. Särkkä. Improved calibration of numerical integration error in sigma-point filters. *arXiv:1811.11474*, 2018.
[51] K. Ritter. *Average Case Analysis of Numerical Problems*. Springer, 2000.
[52] F. Ruggeri, D. R. Insua, and J. Martín. Robust Bayesian analysis. *Handbook of Statistics*, 25:623–667, 2005.
[53] E. G. Ryan, C. C. Drovandi, J. M. McGree, and A. N. Pettitt. A review of modern computational algorithms for Bayesian optimal design. *International Statistical Review*, 84(1):128–154, 2016.
[54] A. V. Sul'din. Wiener measure and its applications to approximation methods. I. *Izvestiâ Vysših Učebnyh Zavedenij Matematika*, 6(13):145–158, 1959.
[55] A. V. Sul'din. Wiener measure and its applications to approximation methods. II. *Izvestiâ Vysših Učebnyh Zavedenij Matematika*, 5(18):165–179, 1960.

[56] G. C. Tiao and B. Afonja. Some Bayesian considerations of the choice of design for ranking, selection and estimation. *Annals of the Institute of Statistical Mathematics*, 28:167–186, 1976.

[57] J. F. Traub. *Information-Based Complexity*. John Wiley and Sons Ltd., 2003.

[58] J. F. Traub and H. Woźniakowski. Information-based complexity: New questions for mathematicians. *Mathematical Intelligencer*, 13(2):34–43, 1991.

[59] A. Tuchscherer. Experimental design for Bayesian estimations in the linear regression model taking costs into account. *Biometrische Zeitschrift*, 25(6):515–525, 1983.

[60] A. Wald. Statistical decision functions which minimize the maximum risk. *Annals of Mathematics*, pages 265–280, 1945.

[61] A. Wald. An essentially complete class of admissible decision functions. *Annals of Mathematical Statistics*, 18(4):549–555, 1947.

[62] J. Wang, J. Cockayne, and C. Oates. On the Bayesian solution of differential equations. In *Proceedings of the 38th International Workshop on Bayesian Inference and Maximum Entropy Methods in Science and Engineering*, 2018.

[63] L. J. Wolfson, J. B. Kadane, and M. J. Small. Expected utility as a policy-making tool: an environmental health example. *Statistics Textbooks and Monographs*, 151:261–278, 1996.

[64] H. Woźniakowski. What is Information-Based Complexity? In *Essays on the Complexity of Continuous Problems*, pages 89–95. European Mathematical Society, 2009.

G. W. Wasilkowski

6 ε-Superposition and truncation dimensions, and multivariate method for ∞-variate linear problems

Abstract: The paper deals with linear problems defined on γ-weighted normed spaces of functions with infinitely many variables. In particular, we present results on ε-truncation and ε-superposition dimensions, which show conditions under which these dimensions are small under modest error demand ε. We also discuss multivariate decomposition methods and their effectiveness in reducing infinitely many variables to a relatively small number equal to the ε-superposition dimension. These positive results are derived for the so called *anchored* decomposition. In the last section, we consider more general decompositions, including the ANOVA one, and show when the ε-truncation/superposition dimensions are small and the multivariate decomposition methods are efficient also for such more general decompositions.

Keywords: Integration, approximation, infinite-variate functions

MSC 2010: 28C20, 41A63, 65J99

6.1 Introduction

This paper deals with the complexity of linear problems defined on weighted spaces of functions with infinitely many variables. Such problems appear in a number of applications, including stochastic differential equations and partial differential equations with random coefficients see, e. g., [22] for a survey.

We follow the *information-based complexity* (IBC) approach by deriving complexity upper and matching lower bounds for the whole space of functions. We also search for algorithms that are efficient for all functions from the space.

Acknowledgement: The paper is based on the invited talk "∞-Variate Linear Problems" at the *Workshop on Multivariate Algorithms and Information-Based Complexity*, Nov. 5–9, 2018, Johann Radon Institute for Computational and Applied Mathematics, Linz, Austria. I would like to thank the organizers for their hospitality and support. I would also like to thank my coauthors: A. Gilbert, M. Gnewuch, M. Hefter, A. Hinrichs, P. Kritzer, F. Y. Kuo, D. Nuyens, F. Pillichshammer, L. Plaskota, K. Ritter, I. H. Sloan, and H. Woźniakowski, for their contribution to some of the results discussed in this paper. Finally, I thank the reviewer of this paper for excellent suggestions how to improve the paper.

G. W. Wasilkowski, Computer Science Department, University of Kentucky, 301 David Marksbury Building, 329 Rose Street, Lexington, KY 40506, USA, e-mail: greg@cs.uky.edu

https://doi.org/10.1515/9783110635461-006

The IBC approach to integration problems for ∞-variate function spaces was initiated in late 90s, see [37], and in the first decade of the 21st century, see [16, 28, 38]. Roughly ten years later, it became a quite popular subject, resulting in a number of important results published in, e. g., [2, 15, 24, 25, 27, 31, 32, 39, 40], and more recently in, e. g., [1, 3–5, 7–12, 14, 17, 19, 21, 23, 33–35].

The majority of papers cited above assume that the space of functions is a reproducing kernel Hilbert space and consider mainly integration problems. In the current paper, we consider spaces that are not necessarily Hilbert and more general linear problems, for which integration problems are just special examples. More specifically, we consider spaces, denoted by $\mathcal{F}_{\gamma}$, of functions f that admit decompositions

$$f = \sum_{\mathfrak{u}} f_{\mathfrak{u}}.$$

Here, the summation is with respect to $\mathfrak{u}$, which are finite subsets of the set $\mathbb{N}$ of positive integers, and each $f_{\mathfrak{u}}$ is a function that depends only on the variables listed in $\mathfrak{u}$, and are members of normed spaces $F_{\mathfrak{u}}$. The space $\mathcal{F}_{\gamma}$ is endowed with a $\boldsymbol{\gamma}$-weighted norm

$$\|f\|_{\mathcal{F}_{\gamma}} = \left[\sum_{\mathfrak{u}} (\gamma_{\mathfrak{u}}^{-1} \|f_{\mathfrak{u}}\|_{F_{\mathfrak{u}}})^p\right]^{1/p},$$

where $\gamma_{\mathfrak{u}}$ are nonnegative numbers, called *weights*, and p is a given number $p \geq 1$ or $p = \infty$. Although some results hold for arbitrary weights $\boldsymbol{\gamma} = (\gamma_{\mathfrak{u}})_{\mathfrak{u}}$, we will concentrate on weights that have *product order-dependent* form

$$\gamma_{\mathfrak{u}} = (|\mathfrak{u}|!)^{\alpha} \prod_{j\in\mathfrak{u}} \gamma_j$$

introduced in [24], or more restrictive *product* form

$$\gamma_{\mathfrak{u}} = \prod_{j\in\mathfrak{u}} \gamma_j$$

introduced in [29]. In both cases, we assume without any loss of generality that

$$\gamma_j \geq \gamma_{j+1} > 0 \quad \text{for all } j \in \mathbb{N}.$$

The problem being considered is the approximation of $\mathcal{S}(f)$, where $\mathcal{S}$ is a linear and continuous operator from $\mathcal{F}_{\gamma}$ into a normed space $\mathcal{G}$,

$$\mathcal{S} : \mathcal{F}_{\gamma} \to \mathcal{G}.$$

We discuss two approaches to deal with such problems with ∞-variate functions. The first approach is to reduce the infinite number of variables to a finite (and possibly small) number k of variables. This leads to the concept of *ε-truncation dimension* considered in Section 6.3. Roughly speaking, for a given error demand ε, the ε-truncation dimension (denoted by $\dim^{\mathrm{trnc}}(\varepsilon, \mathcal{S})$) is the smallest number k such that

$$\|\mathcal{S}(f) - \mathcal{S}(f_k)\|_{\mathcal{G}} \le \varepsilon \, \|f - f_k\|_{\mathcal{F}_\gamma} \quad \text{for all } f \in \mathcal{F}_\gamma.$$

Here, f_k is a k-variate function obtained from f by

$$f_k(x_1, \dots, x_k) = f(x_1, \dots, x_k, 0, 0, \dots).$$

We stress that unlike in financial mathematics and statistics, our concept of ε-truncation dimension is defined in the worst-case setting with respect to the whole space $\mathcal{F}_\gamma$. This approach has been first considered in [19], see also [17, 21]. The presentation in Section 6.3 is based on [17] that contains most general results up to now. It turns out that for product weights and *anchored* decompositions, the ε-truncation dimension increases slowly with decreasing ε. By anchored decomposition, we mean that each $f_{\mathfrak{u}}(\boldsymbol{x})$ is zero if at least one coordinate $x_j = 0$ with $j \in \mathfrak{u}$. For nonanchored decompositions, we show that, the so called, *ε-pseudo truncation dimension* has the same properties, i. e., is increasing slowly with $\varepsilon \to 0$. See also the last Section 6.6 and the Corollary 6.16.

The second approach to reduce the infinitely many variables is discussed in Sections 6.4 and 6.5. In the former section, we consider the *ε-superposition dimension*, denoted by $\dim^{\mathrm{sprp}}(\varepsilon, \mathcal{S})$. Roughly speaking, it is the smallest number k such that, for every $f \in \mathcal{F}_\gamma$, $\mathcal{S}(f)$ can be approximated (with the worst-case error $\le \varepsilon$) by a sum $\mathcal{S}(\sum_{\mathfrak{u} \in \mathcal{V}_\varepsilon} f_{\mathfrak{u}})$, where each $f_{\mathfrak{u}}$ in the sum depends only on at most k variables, i. e., $k = \max_{\mathfrak{u} \in \mathcal{V}_\varepsilon} |\mathfrak{u}|$. Moreover, $\mathcal{V}_\varepsilon$ does not depend on specific functions f. As for the truncation dimension, the concept of ε-superposition dimension is different than the one in finance or statistics. The presentation in the section is based on [8, 27]. Under suitable assumptions, it was shown in [27] that the superposition dimension is as small as

$$\dim^{\mathrm{sprp}}(\varepsilon, \mathcal{S}) = O\left(\frac{\ln(1/\varepsilon)}{\ln(\ln(1/\varepsilon))}\right) \quad \text{as } \varepsilon \to 0,$$

and that the corresponding set $\mathcal{V}_\varepsilon$ is relatively small. In [8], we provide an efficient algorithm for constructing the smallest possibly $\mathcal{V}_\varepsilon$ for product weights. The concept of ε-superposition dimension, and the fact that it is small, has been used earlier (without the official name) in a number of papers studying *multivariate decomposition methods* (MDM), discussed in Section 6.5.

MDM were first introduced in [25], under the name of *changing dimension algorithms*, and then investigated in a number of papers, including [4, 7, 8, 11, 23, 27, 31–35, 39, 40]. The main idea behind MDM is to approximate $\mathcal{S}(f_{\mathfrak{u}})$ for $\mathfrak{u} \in \mathcal{V}_\varepsilon$, where $\mathcal{V}_\varepsilon$ is the set from the definition of the ε-superposition dimension. To apply algorithms to specific terms $f_{\mathfrak{u}}$, we assume that the decomposition is anchored. Then, for product order-dependent weights, the corresponding MDM has the worst-case error $\le \varepsilon$ at the cost bounded from above by

$$O(\varepsilon^{-\kappa-\delta}) \quad \text{for all } \delta > 0, \quad \text{where} \quad \kappa = \max\left(\frac{1}{r}, \frac{1}{\beta - 1/p^*}\right).$$

Here, $1/r$ is the optimal rate for the univariate problem, β the order of convergence of the weights $\gamma_{\mathfrak{u}}$, and p^* the conjugate of p. As follows from a lower bound in [25], this is the optimal rate (modulo δ). In particular, for β sufficiently large, the rate κ is the same as for the corresponding problem with univariate functions only. We stress that these positive results hold in a very demanding cost model from [25], where a function evaluation can be performed at points $\boldsymbol{x}$ with only finite number, say k, of nonzero coordinates x_j, and at the cost of $\$(k)$, where the cost function $\$$ can be from linear to exponential in k,

$$\Omega(k) = \$(k) = \exp(O(k)) \quad \text{as } k \to \infty.$$

The main results discussed so far are derived under the anchored decomposition assumptions, since, as shown in [26], it is possible to compute samples of the unknown terms $f_{\mathfrak{u}}$ using samples of the whole function $f = \sum_{\mathfrak{u}} f_{\mathfrak{u}}$. Such property does not hold for many other decompositions, including the famous ANOVA one. In particular, it is not clear if the functions f_k (from the ε-truncation dimension definition) belong to the corresponding space and if $\mathcal{S}(f_k)$ are well defined.

However, small ε-truncation and superposition dimensions, as well as efficiency of MDM's hold also for other decompositions if the corresponding anchored and unanchored decomposition norms are equivalent. This is why the final section, Section 6.6, deals with the equivalence of norms under different decompositions. Research on this subject has been initiated in a seminal paper [13] by M. Hefter and K. Ritter, where the equivalence between anchored and ANOVA norms in a Hilbert space setting with product weights was shown. Later, this result was extended in [14, 18–20] to non-Hilbert spaces and general weights. Our presentation in Section 6.6 is based on a recent paper [17], which contains the most general results up to now.

6.2 Basic definitions

6.2.1 Building intuition

To explain various concepts and assumptions, we consider in this section the following example that is a very special case of the more general problems discussed in the current paper.

Consider a smooth function

$$g : \mathbb{R} \to \mathbb{R},$$

and a random process

$$\mathbb{X}(t) = \sum_{j=1}^{\infty} x_i\, \xi_j(t),$$

where x_j's are i. i. d. (independent and identically distributed) random variables with the corresponding probability measure μ, and $\xi_j(t)$ converging to 0 sufficiently fast. Consider next the problem of approximating the expectation of $g(\mathbb{X}(t_0))$ for a given and fixed time t_0. This is equivalent to approximating the following ∞-variate integral:

$$\int_{\mathbb{R}^{\mathbb{N}}} f(\boldsymbol{x})\, \mu^{\mathbb{N}}(\,\mathrm{d}\boldsymbol{x}) \quad \text{for a function} \quad f(\boldsymbol{x}) := g\left(\sum_{j=1}^{\infty} x_j\, \xi(t_0) \right),$$

with respect to the corresponding countable product measure $\mu^{\mathbb{N}}$. This motivates the study of integration with ∞-many variables.

Using, e. g., Taylor's series expansion,

$$f(\boldsymbol{x}) = \sum_{\mathfrak{u}} f_{\mathfrak{u}},$$

where $\mathfrak{u}$'s are finite subsets of positive integers, and each $f_{\mathfrak{u}}$ depends only on variables x_j with $j \in \mathfrak{u}$. For instance, for $\mathfrak{u}$ equal to $\emptyset$, $\{1\}$, or $\{2,5\}$, we respectively have

$$f_{\emptyset}(\boldsymbol{x}) \equiv g(0), \quad f_{\{1\}}(x_1) = \sum_{j=1}^{\infty} \frac{\xi_1^j(t_0)}{j!}\, g^{(j)}(0)\, x_1^j,$$

and

$$f_{2,5}(x_2, x_5) = \sum_{i,j=1}^{\infty} \frac{\xi_2^i(t_0)\, \xi_5^j(t_0)}{i!\, j!}\, g^{(i+j)}(0)\, x_2^i\, x_5^j.$$

This is a special case of decompositions used in the paper. Actually, since

$$f_{\{1\}}(0) = f_{\{2,5\}}(0, x_5) = f_{\{2,5\}}(x_2, 0) = 0,$$

this is an example of the so called *anchored decomposition.* Note also that the terms depend on $\xi_j(t_0)$ that converge to zero. The products of $|\xi_j(t_0)|$ are related to weights and weighted spaces considered in the paper.

Finally, note that evaluation of f at an arbitrary point $\boldsymbol{x} = (x_1, x_2, \dots)$ would require infinitely many arithmetic operations. This explains why we consider here evaluation of f at points $\boldsymbol{x} = (x_1, x_2, \dots)$ with only finitely many, say k, nonzero coordinates x_j, and the cost of such an evaluation has to be at least linear in k. For instance, if all x_j, but x_1, x_3, x_7, are zero, then computing $f(\boldsymbol{x}) = g(x_1\, \xi_1(t_0) + x_3\, \xi_3(t_0) + x_7\, \xi_7(t_0))$ should cost at least $3 \cdot$ constant.

6.2.2 Notation

We begin by introducing the notation used in this paper. By $\mathbb{N}$, we denote the set of all positive integers, and by $\mathcal{U}$ the collection of all finite and increasing sequences $\mathfrak{u}$ of positive integers of the form

$$\mathfrak{u} = (u_1, \dots, u_k), \quad \text{where} \quad u_i \in \mathbb{N} \quad \text{and} \quad u_j < u_{j+1}.$$

Of course, $\emptyset$ denotes the empty sequence, i. e., $k = 0$. Sometimes it is convenient to treat $\mathfrak{u}$ as a set and, in particular, perform set operations on them, write $k = |\mathfrak{u}|$, or

$$j \in \mathfrak{u} \quad \text{iff} \quad \text{there is } \ell \le |\mathfrak{u}| \text{ such that } j = u_\ell.$$

Let D be a nonempty interval in $\mathbb{R}$ (possibly infinite like $\mathbb{R}_+ = [0, \infty)$ or $\mathbb{R} = (-\infty, \infty)$), and let $D^{\mathbb{N}}$ be the set of infinite sequences $\boldsymbol{x} = (x_j)_{j \in \mathbb{N}}$ (sometimes referred to as *points*) with $x_j \in D$. Without any loss of generality, we assume that $0 \in D$ so that $\mathbf{0} = (0, 0, \dots) \in D^{\mathbb{N}}$.

For given $\mathfrak{u} = (u_1, \dots, u_k) \in \mathcal{U}$ with $k = |\mathfrak{u}|$ and $\boldsymbol{x} \in D^{\mathbb{N}}$, we define

$$\boldsymbol{x}_\mathfrak{u} = (x_{u_1}, \dots, x_{u_k}) \quad \text{and} \quad [\boldsymbol{x}; \mathfrak{u}] = \boldsymbol{y} \quad \text{with} \quad y_j = \begin{cases} x_j & \text{if } j \in \mathfrak{u}, \\ 0 & \text{if } j \notin \mathfrak{u}. \end{cases}$$

In particular, $[\boldsymbol{x}; \emptyset] = \mathbf{0} = (0, 0, \dots)$. In a number of papers, $[\boldsymbol{x}; \mathfrak{u}]$ is denoted by a less formal notation $[\boldsymbol{x}_\mathfrak{u}; \mathbf{0}]$.

6.2.3 γ-Weighted spaces $\mathcal{F}_\gamma$

For every $\mathfrak{u} \in \mathcal{U}$, let $F_\mathfrak{u}$ be a normed space of functions

$$f_\mathfrak{u} : D^{\mathbb{N}} \to \mathbb{R} \quad \text{that depend only on } \boldsymbol{x}_\mathfrak{u}$$

with the norm denoted by $\| \cdot \|_{F_\mathfrak{u}}$. In particular, $F_\emptyset$ is the space of constant functions with the absolute value as its norm. To stress that $f_\mathfrak{u}$ depends only on the variables listed in $\mathfrak{u}$, we will write sometimes $f_\mathfrak{u}(\boldsymbol{x}) = f_\mathfrak{u}(\boldsymbol{x}_\mathfrak{u})$. We assume that

$$F_\mathfrak{u} \cap F_\mathfrak{v} = \{0\} \quad \text{if} \quad \mathfrak{u} \neq \mathfrak{v}.$$

Let $\boldsymbol{\gamma} = (\gamma_\mathfrak{u})_{\mathfrak{u} \in \mathcal{U}}$ be a given family of nonnegative numbers $\gamma_\mathfrak{u}$, called *weights*, and

$$\mathcal{U}_+ = \{\mathfrak{u} \in \mathcal{U} : \gamma_\mathfrak{u} > 0\},$$

and let $F_\infty = \operatorname{span}\{F_{\mathfrak{u}} : \mathfrak{u} \in \mathcal{U}_+\}$ denote the linear space of finite linear combinations of functions $f_{\mathfrak{u}}$. Then $\mathcal{F}_\gamma$ is the completion[1] of F_∞ with respect to the norm

$$\|f\|_{\mathcal{F}_\gamma} = \left[\sum_{\mathfrak{u}\in\mathcal{U}_+} \gamma_{\mathfrak{u}}^{-p} \|f_{\mathfrak{u}}\|_{F_{\mathfrak{u}}}^p\right]^{1/p} < \infty, \tag{6.1}$$

where

$$f = \sum_{\mathfrak{u}\in\mathcal{U}_+} f_{\mathfrak{u}} \quad \text{with} \quad f_{\mathfrak{u}} \in F_{\mathfrak{u}} \tag{6.2}$$

is the unique decomposition of $f \in \mathcal{F}_\gamma$. Here $p \in [1,\infty]$, with

$$\|f\|_{\mathcal{F}_\gamma} = \sup_{\mathfrak{u}\in\mathcal{U}_+} \gamma_{\mathfrak{u}}^{-1} \|f_{\mathfrak{u}}\|_{F_{\mathfrak{u}}} \quad \text{for } p = \infty.$$

To stress the dependence on p, we will write sometimes $\mathcal{F}_{\gamma,p}$ instead of $\mathcal{F}_\gamma$.

We will say that $\mathcal{F}_\gamma$ is an *anchored* space if for every nonempty $\mathfrak{u}$ and any $f_{\mathfrak{u}} \in F_{\mathfrak{u}}$ we have

$$f_{\mathfrak{u}}(\boldsymbol{x}) = 0 \quad \text{if } x_j = 0 \text{ for some } j \in \mathfrak{u}. \tag{6.3}$$

Remark 6.1. For some specific spaces $F_{\mathfrak{u}}$, there might exist functions $f_{\mathfrak{u}} \in F_{\mathfrak{u}}$ and points $\boldsymbol{x} \in D^{\mathbb{N}}$ such that the series in (6.2) is not convergent pointwise, i. e., $\sum_{\mathfrak{u}\in\mathcal{U}_+} |f_{\mathfrak{u}}(\boldsymbol{x}_{\mathfrak{u}})| = \infty$. In such cases, we will treat $\mathcal{F}_\gamma$ as a space of sequences $f = (f_{\mathfrak{u}})_{\mathfrak{u}\in\mathcal{U}_+}$. Note however, that for anchored spaces, $f([\boldsymbol{x};\mathfrak{v}])$ is always well defined for any $\boldsymbol{x} \in D^{\mathbb{N}}$ and $\mathfrak{v} \in \mathcal{U}_+$, since then $f([\boldsymbol{x};\mathfrak{v}]) = \sum_{\mathfrak{u}\subseteq\mathfrak{v}} f_{\mathfrak{u}}(\boldsymbol{x}_{\mathfrak{u}})$ is a finite sum.

Remark 6.2. The results discussed in this paper are also applicable to functions with a finite number, denoted by d, of variables. In such a case, we have to replace $\mathcal{U}$ by $\mathcal{U}_d = \{\mathfrak{u} : \mathfrak{u} \subseteq [1,\dots,d]\}$ or, equivalently, set $\gamma_{\mathfrak{u}} = 0$ for all $\mathfrak{u} \not\subseteq [1,\dots,d]$.

6.2.4 Specific weights

Although we consider arbitrary weights, main results presented in this paper assume either *product weights* of the form

$$\gamma_{\mathfrak{u}} = \prod_{j\in\mathfrak{u}} \gamma_j$$

introduced in [29], or *product order-dependent weights* of the form

$$\gamma_{\mathfrak{u}} = (|\mathfrak{u}|!)^\alpha \prod_{j\in\mathfrak{u}} \gamma_j$$

1 The results presented in this paper do not require the completion, i. e., they hold for the space F_∞ with norm (6.1).

introduced in [24]. In both cases γ_j are positive numbers and, without any loss of generality $\gamma_j \geq \gamma_{j+1}$ for all $j \in \mathbb{N}$.

For given weights, the *decay* is defined as

$$\mathrm{decay}(\boldsymbol{\gamma}) := \sup\Big(\tau : \sum_{\mathfrak{u}\in\mathcal{U}} \gamma_{\mathfrak{u}}^{1/\tau} < \infty\Big).$$

For $\gamma_j = j^{-\beta}$, the decay of the corresponding product weights is β. It is also β for the corresponding product order-dependent weights assuming that $\alpha < \beta$.

6.2.5 Linear problem

Let $\mathcal{G}$ be a normed space, whose norm is denoted by $\|\cdot\|_{\mathcal{G}}$. Let $\mathcal{S}$ be a linear operator

$$\mathcal{S} : \mathcal{F}_{\gamma,p} \to \mathcal{G},$$

and let

$$S_{\mathfrak{u}} = \mathcal{S}|_{F_{\mathfrak{u}}} : F_{\mathfrak{u}} \to \mathcal{G}$$

be the restrictions of $\mathcal{S}$ to the spaces $F_{\mathfrak{u}}$. Let $\|S_{\mathfrak{u}}\|$ denote the corresponding operator norms,

$$\|S_{\mathfrak{u}}\| = \sup_{f_{\mathfrak{u}}\in F_{\mathfrak{u}}} \frac{\|S_{\mathfrak{u}}(f_{\mathfrak{u}})\|_{\mathcal{G}}}{\|f_{\mathfrak{u}}\|_{F_{\mathfrak{u}}}}.$$

Through this paper p^* is the conjugate of p, i. e.,

$$\frac{1}{p} + \frac{1}{p^*} = 1.$$

We assume that

$$\Big[\sum_{\mathfrak{u}\in\mathcal{U}} (\gamma_{\mathfrak{u}}\,\|S_{\mathfrak{u}}\|)^{p^*}\Big]^{1/p^*} < \infty, \tag{6.4}$$

with

$$\Big[\sum_{\mathfrak{u}\in\mathcal{U}} (\gamma_{\mathfrak{u}}\,\|S_{\mathfrak{u}}\|)^{p^*}\Big]^{1/p^*} = \sup_{\mathfrak{u}\in\mathcal{U}} \gamma_{\mathfrak{u}}\,\|S_{\mathfrak{u}}\| \quad \text{for} \quad p^* = \infty.$$

This is because (6.4) implies continuity of $\mathcal{S}$. Indeed, we have the following proposition:

Proposition 6.3. *The operator norm of $\mathcal{S}$ is bounded by*

$$\|\mathcal{S}\| \le \left[\sum_{\mathfrak{u}\in\mathcal{U}} (\gamma_{\mathfrak{u}} \|S_{\mathfrak{u}}\|)^{p^*}\right]^{1/p^*}$$

with equality if $\mathcal{S}$ is a functional or $p^* = \infty$.

Proof. For any $f = \sum_{\mathfrak{u}\in\mathcal{U}_+} f_{\mathfrak{u}}$, we have

$$\|\mathcal{S}(f)\|_{\mathcal{G}} \le \sum_{\mathfrak{u}\in\mathcal{U}_+} \|S_{\mathfrak{u}}\| \, \|f_{\mathfrak{u}}\|_{F_{\mathfrak{u}}} = \sum_{\mathfrak{u}\in\mathcal{U}_+} \gamma_{\mathfrak{u}} \|S_{\mathfrak{u}}\| \, \gamma_{\mathfrak{u}}^{-1} \|f_{\mathfrak{u}}\|_{F_{\mathfrak{u}}} \le \left[\sum_{\mathfrak{u}\in\mathcal{U}} (\gamma \|S_{\mathfrak{u}}\|)^{p^*}\right]^{1/p^*} \|f\|_{\mathcal{F}_{\gamma,p}}.$$

This proves the upper bound on $\|\mathcal{S}\|$. The equality (when $\mathcal{S}$ is a functional or $p^* = \infty$) follows from the facts that the first inequality is sharp for f_u such that $\|S_{\mathfrak{u}}(f_{\mathfrak{u}})\| = \|S_{\mathfrak{u}}\| \, \|f_{\mathfrak{u}}\|_{F_{\mathfrak{u}}}$ and Hölder's inequality is sharp. □

If the weights have product form and $\mathcal{S}$ satisfies the following condition that there exists a constant C_1 such that

$$\|S_{\mathfrak{u}}\| \le C_1^{|\mathfrak{u}|} \quad \text{for all } \mathfrak{u} \in \mathcal{U}_+, \tag{6.5}$$

then

$$\left[\sum_{\mathfrak{u}\in\mathcal{U}} (\gamma_{\mathfrak{u}} \|S_{\mathfrak{u}}\|)^{p^*}\right]^{1/p^*} \le \left[\prod_{j=1}^{\infty} (1 + (\gamma_j \, C_1)^{p^*})\right]^{1/p*}$$

for $p^* < \infty$, and

$$\left[\sum_{\mathfrak{u}\in\mathcal{U}} (\gamma_{\mathfrak{u}} \|S_{\mathfrak{u}}\|)^{p^*}\right]^{1/p^*} \le \sup_k C_1^k \prod_{j=1}^{k} \gamma_j$$

for $p^* = \infty$ (recall that $\gamma_j \ge \gamma_{j+1}$).

We will illustrate some of the results using the following integration problem:

Integration example. Let $D = [0,1]$ and $F_{\mathfrak{u}}$ consist of

$$f_{\mathfrak{u}}(\boldsymbol{x}_{\mathfrak{u}}) = \int_{D^{|\mathfrak{u}|}} \prod_{j\in\mathfrak{u}} (x_j - t_j)_+^0 \, h(\boldsymbol{t}_{\mathfrak{u}}) \, \mathrm{d}\boldsymbol{t}_{\mathfrak{u}} \quad \text{for} \quad h_{\mathfrak{u}} \in L_p(D^{|\mathfrak{u}|})$$

with the norm equal to the L_p norm of $h_{\mathfrak{u}}$, i. e., the L_p norm, the mixed partial first-order derivatives of $f_{\mathfrak{u}}$. Let

$$S_{\mathfrak{u}}(f_{\mathfrak{u}}) = \int_{[0,1]^{|\mathfrak{u}|}} f_{\mathfrak{u}}(\boldsymbol{x}_{\mathfrak{u}}) \, \mathrm{d}\boldsymbol{x}_{\mathfrak{u}} \quad \text{with} \quad S_{\emptyset}(f_{\emptyset}) = f_{\emptyset},$$

then

$$\|S_{\mathfrak{u}}\| = (1 + p^*)^{-|\mathfrak{u}|/p^*} \quad \text{with } \|S_{\mathfrak{u}}\| = 1 \text{ for } p^* = \infty.$$

6.3 ε-Truncation dimension

The ε-truncation dimension, introduced in [19], depends on the error demand ε and the problem $\mathcal{S}$, and is defined as follows: For a natural number $k \in \{0\} \cup \mathbb{N}$ and a function $f \in \mathcal{F}_\gamma$, consider

$$f_k(\boldsymbol{x}) := f(x_1, \dots, x_k, 0, 0 \dots) = f([\boldsymbol{x}; \{1, \dots, k\}]).$$

Definition 6.4. The *ε-truncation dimension* is

$$\dim^{\mathrm{trnc}}(\varepsilon, \mathcal{S}) := \min\left(k : \sup_{f \in \mathcal{F}_\gamma} \frac{\|\mathcal{S}(f) - \mathcal{S}(f_k)\|_{\mathcal{G}}}{\|f - f_k\|_{\mathcal{F}_\gamma}} \le \varepsilon \right).$$

Note that the dimension is defined in the worst-case sense[2] with respect to the whole space $\mathcal{F}_\gamma$. Hence, it differs from the *effective truncation dimension* in statistics that depends on individual functions.

Definition 6.4 implicitly assumes that $f_k \in \mathcal{F}_\gamma$, so that $\mathcal{S}(f_k)$ is well defined. However, this is not the case in general (see Definition 6.6 and the last section).

We have the following result that is a simple extension of results from [17, 19]:

Theorem 6.5. *Assume that $\mathcal{F}_{\gamma,p}$ is an anchored space. Then*

$$\dim^{\mathrm{trnc}}(\varepsilon, \mathcal{S}) \le \min\{k : \mathrm{er}(k) \le \varepsilon\},$$

where

$$\mathrm{er}(k) := \begin{cases} [\sum_{\mathfrak{u} \not\subseteq \{1,\dots,k\}} (\gamma_{\mathfrak{u}} \|S_{\mathfrak{u}}\|)^{p^*}]^{p^*} & \text{if } p^* < \infty, \\ \sup_{\mathfrak{u} \not\subseteq \{1,\dots,k\}} \gamma_{\mathfrak{u}} \|S_{\mathfrak{u}}\| & \text{if } p^* = \infty. \end{cases}$$

Assume additionally that the weights have product form, $\gamma_{\mathfrak{u}} = \prod_{j \in \mathfrak{u}} \gamma_j$, and $\mathcal{S}$ satisfies (6.5). Then

$$\mathrm{er}(k) \le \left[\prod_{j=1}^{\infty} \left(1 + (\gamma_j C_1)^{p^*}\right) \right]^{1/p^*} \left(1 - \exp\left(- \sum_{j=k+1} (\gamma_j C_1)^{p^*} \right)\right)^{1/p^*} \quad \text{if } p^* < \infty,$$

and

$$\mathrm{er}(k) \le \sum_{\ell \ge k+1} C_1^\ell \prod_{j=1}^{\ell} \gamma_j \quad \text{if } p^* = \infty.$$

---

2 ε-truncation dimension has also been defined and studied in the *average case setting*, with respect to a probability measure on $\mathcal{F}_\gamma$, see [21].

We now explain the importance of the low ε-truncation dimension. Let $F_{k,p}$ be the subspace of $\mathcal{F}_{\gamma,p}$ that is spanned by F_{u} with $\mathrm{u} \subseteq \{1,\ldots,k\}$. Of course this space consists of functions f_k that depend on the first k variables. Let $A_{\varepsilon,k}(f_k)$ be an algorithm for approximating $\mathcal{S}(f_k)$ with the worst-case error (with respect to $F_{k,p}$ space) equal ε,

$$\|\mathcal{S}(f_k) - A_{\varepsilon,k}(f_k)\|_{\mathcal{G}} \le \varepsilon \|f_k\|_{\mathcal{F}_{\gamma,p}}.$$

Then for every $\tau \in (0,1)$, the algorithm $\mathcal{A}_{\varepsilon,\tau,p}$,

$$\mathcal{A}_{\varepsilon,\tau,p}(f) = A_{\eta,k}(f_k) \quad \text{with} \quad k = \dim^{\mathrm{trnc}}(\varepsilon\,(1-\tau^{p^*})^{1/p^*}) \quad \text{and} \quad \eta = \varepsilon\cdot\tau,$$

has the worst-case error (with respect to the whole space $\mathcal{F}_{\gamma,p}$) bounded by ε. Of course, $(1-\tau^{p^*})^{1/p^*} = 1$ if $p^* = 1$. In particular, for $\tau = 1/2^{1/p^*}$, we have $(1-\tau^{p^*})^{1/p^*} = \tau$.

We now provide vales of the ε-truncation dimension for the

Integration example (continued). Consider $\varepsilon = 10^{-3}$ (for other values of ε see [19]) and $\gamma_j = j^{-\beta}$. Then the values of 10^{-3}-truncation dimension are

31 for $\beta = 2$, 9 for $\beta = 3$, 5 for $\beta = 4$ for $p = 1$, and
68 for $\beta = 2$, 11 for $\beta = 3$, 5 for $\beta = 4$ for $p = 2$.

The proof in [19] uses the fact that for anchored spaces $\mathcal{F}_{\gamma,p}$,

$$f_k(\boldsymbol{x}) = f([\boldsymbol{x};\{1,\ldots,k\}]) = \sum_{\mathrm{u}\subseteq\{1,\ldots,k\}} f_{\mathrm{u}}(\boldsymbol{x}),$$

and therefore

$$\mathcal{S}(f) - \mathcal{S}(f_k) = \sum_{\mathrm{u}\not\subseteq\{1,\ldots,k\}} S_{\mathrm{u}}(f_{\mathrm{u}}).$$

The rest of the proof is to estimate $\sum_{\mathrm{u}\not\subseteq\{1,\ldots,k\}} \|S_{\mathrm{u}}(f_{\mathrm{u}})\|_{\mathcal{G}}$. This implies that for general (not necessarily anchored) weighted spaces $\mathcal{F}_{\gamma,p}$, we have the same upper bounds as in Theorem 6.5, for the extended notion of ε-truncation dimension, whose formal definition is as follows:

Definition 6.6. For any decomposition, the *ε-pseudo-truncation dimension* is

$$\dim^{\text{p-trnc}}(\varepsilon,\mathcal{S}) := \min\left(k : \sup_{f\in\mathcal{F}_{\gamma,p}} \frac{\|\sum_{\mathrm{u}\not\subseteq\{1,\ldots,k\}} S_{\mathrm{u}}(f_{\mathrm{u}})\|_{\mathcal{G}}}{\|\sum_{\mathrm{u}\not\subseteq\{1,\ldots,k\}} f_{\mathrm{u}}\|_{\mathcal{F}_{\gamma,p}}} \le \varepsilon\right).$$

6.4 ε-Superposition dimension

We begin with the following definition:

Definition 6.7. For given error demand $\varepsilon > 0$, we say that a subset $\mathcal{V}$ of $\mathcal{U}_+$ is *ε-active* iff

$$\sup_{f\in\mathcal{F}_\gamma} \frac{\|\sum_{\mathfrak{u}\in\mathcal{U}_+\setminus\mathcal{V}} S_{\mathfrak{u}}(f_{\mathfrak{u}})\|_{\mathcal{G}}}{\|\sum_{\mathfrak{u}\in\mathcal{U}_+\setminus\mathcal{V}} f_{\mathfrak{u}}\|_{\mathcal{F}_\gamma}} \le \varepsilon.$$

Then the *ε-superposition dimension* is

$$\dim^{\mathrm{sprp}}(\varepsilon, \mathcal{S}) := \inf\{\|\mathcal{V}\|_\infty : \mathcal{V} \text{ is } \varepsilon\text{-active}\}, \quad \text{where} \quad \|\mathcal{V}\|_\infty := \sup_{\mathfrak{u}\in\mathcal{V}} |\mathfrak{u}|. \tag{6.6}$$

Sometimes we will simply write $\dim^{\mathrm{sprp}}(\varepsilon)$. The following theorem provides sharp results for $p = 1$ and upper bounds on the dimension for $p > 1$.

Theorem 6.8. *Suppose that*

$$\tau_0 := \mathrm{decay}((\gamma_{\mathfrak{u}}\,\|S_{\mathfrak{u}}\|)_{\mathfrak{u}}) > 1.$$

(i) *Consider $p = 1$. Then*

$$\dim^{\mathrm{sprp}}(\varepsilon, \mathcal{S}) = \|\mathcal{V}_{\varepsilon,1}\|_\infty, \quad \textit{where} \quad \mathcal{V}_{\varepsilon,1} = \{\mathfrak{u} : \gamma_{\mathfrak{u}}\,\|S_{\mathfrak{u}}\| > \varepsilon\}.$$

Moreover, $\mathcal{V}_{\varepsilon,1}$ is the smallest ε-active set, i. e., it is contained in any ε-active set, and its cardinality is bounded by

$$|\mathcal{V}_{\varepsilon,1}| < \frac{\sum_{\mathfrak{u}\in\mathcal{U}_+} (\gamma_{\mathfrak{u}}\,\|S_{\mathfrak{u}}\|)^{(1+\delta)/\tau_0}}{\varepsilon^{(1+\delta)/\tau_0}} \quad \textit{for any } \delta > 0.$$

(ii) *Consider $p > 1$. Then for any $a \in (1/(\tau_0\, p^*), 1)$,*

$$\dim^{\mathrm{sprp}}(\varepsilon, \mathcal{S}) \le \|\mathcal{V}_{\varepsilon,p,a}\|_\infty,$$

where

$$\mathcal{V}_{\varepsilon,p,a} = \left\{\mathfrak{u} : (\gamma_{\mathfrak{u}}\,\|S_{\mathfrak{u}}\|)^{1-a} > \frac{\varepsilon}{(\sum_{\mathfrak{v}\in\mathcal{U}_+} (\gamma_{\mathfrak{v}}\,\|S_{\mathfrak{v}}\|)^{a p^*})^{1/p^*}}\right\}.$$

Moreover, each $\mathcal{V}_{\varepsilon,p,a}$ is ε-active with the cardinality bounded by

$$|\mathcal{V}_{\varepsilon,p,a}| \le \frac{(\sum_{\mathfrak{v}\in\mathcal{U}_+} (\gamma_{\mathfrak{v}}\,\|S_{\mathfrak{v}}\|)^{(1+\delta)/\tau_0})\,(\sum_{\mathfrak{v}\in\mathcal{U}_+} (\gamma_{\mathfrak{v}}\,\|S_{\mathfrak{v}}\|)^{a p^*})^{(1+\delta)/((1-a)\,\tau_0\, p^*)}}{\varepsilon^{(1+\delta)/((1-a)\,\tau_0)}}.$$

Proof. For $p = 1$ and any $\mathcal{V}$,

$$\Big\|\sum_{\mathfrak{v}\notin\mathcal{V}} S(f_{\mathfrak{v}})\Big\|_{\mathcal{G}} \le \Big\|\sum_{\mathfrak{v}\notin\mathcal{V}} f_{\mathfrak{v}}\Big\|_{\mathcal{F}_{\gamma,1}} \sup_{\mathfrak{u}\notin\mathcal{V}} \gamma_{\mathfrak{u}}\,\|S_{\mathfrak{u}}\|,$$

which proves that $\mathcal{V}_{\varepsilon,1}$ is ε-active. To prove that it is the smallest ε-active set, consider $\mathfrak{u} \in \mathcal{V}_{\varepsilon,1}$ and $f = f_u^* \neq 0$, for which $\|S_\mathfrak{u}(f_\mathfrak{u}^*)\|_\mathcal{G} = \|S_\mathfrak{u}\| \, \|f_\mathfrak{u}^*\|_{F_\mathfrak{u}}$. Then

$$\|\mathcal{S}(f_\mathfrak{u}^*)\|_\mathcal{G} = \gamma_\mathfrak{u} \|S_\mathfrak{u}\| \, \|f_\mathfrak{u}^*\|_{\mathcal{F}_{\gamma,p}} > \varepsilon \, \|f_\mathfrak{u}^*\|_{\mathcal{F}_{\gamma,p}},$$

which implies that any ε-active set $\mathcal{V}_\varepsilon$ has to contain $\mathcal{V}_{\varepsilon,1}$. To estimate the cardinality of $\mathcal{V}_{\varepsilon,1}$, let $(\lambda_j)_{j\in\mathbb{N}}$ be a sequence of all $\gamma_\mathfrak{u} \|S_\mathfrak{u}\|$ rearranged in the decreasing order. Since $\sum_{j=1}^\infty \lambda_j^{(1+\delta)/\tau_0}$ is finite, we see that the kth largest value is

$$\lambda_k \leq \left(\frac{\sum_{\mathfrak{v}\in\mathcal{U}_+} (\gamma_\mathfrak{v} \|S_\mathfrak{v}\|)^{(1+\delta)/\tau_0}}{k} \right)^{\tau_0/(1+\delta)},$$

which completes the proof of part (i).

The part (ii) can be shown in a similar way using the fact that for any $\mathcal{V} \subseteq \mathcal{U}_+$, we have

$$\sum_{\mathfrak{u}\notin\mathcal{V}} \|S_\mathfrak{u}(f_\mathfrak{u})\|_\mathcal{G} \leq \left[\sum_{\mathfrak{u}\notin\mathcal{V}} \gamma_\mathfrak{u}^{-p} \|f_\mathfrak{u}\|_{F_\mathfrak{u}}^p \right]^{1/p} \left[\max_{\mathfrak{v}\notin\mathcal{V}} (\gamma_\mathfrak{v} \|S_\mathfrak{v}\|)^{(1-a)p^*} \sum_{\mathfrak{u}\in\mathcal{U}_+} (\gamma_\mathfrak{u} \|S_\mathfrak{u}\|)^{ap^*} \right]^{1/p^*},$$

which completes the proof. □

As in [27], see also [23, 25, 31–34], we have the following corollary:

Corollary 6.9. *If* (6.5) *holds and* $\gamma_\mathfrak{u} = C(|\mathfrak{u}|!)^\alpha \prod_{j\in\mathfrak{u}} c/j^\beta$ *with* $\beta > \max(1, \alpha)$, *then*

$$\dim^{\mathrm{sprp}}(\varepsilon) = O\left(\frac{\ln(1/\varepsilon)}{\ln(\ln(1/\varepsilon))} \right) \quad \text{as } \varepsilon \to 0.$$

Remark 6.10. For product weights and any $p \in [1, \infty]$, there is in [8] a very efficient algorithm for constructing an ε-active set $\mathcal{V}_{\varepsilon,p}$ that attains the infimum in (6.6), and has the smallest cardinality among all ε-active sets. To stress its optimality, we will denote it by

$$\mathrm{Act}(\varepsilon, p) = \mathcal{V}_{\varepsilon,p}.$$

Of course, it coincides with $\mathcal{V}_{\varepsilon,1}$ if $p = 1$. Hence, $\|\mathrm{Act}(\varepsilon, p)\|_\infty$ provides the exact value of $\dim^{\mathrm{sprp}}(\varepsilon, \mathcal{S})$ at least while $\mathcal{S}$ is a functional.

We now provide values of the ε-superposition dimension and the cardinality of $\mathrm{Act}(\varepsilon, p)$ for the integration example. For more see [8].

Integration example (continued). Consider $\varepsilon = 10^{-3}$ and product weights with $\gamma_j = j^{-\beta}$. Then the ε-superposition dimension and the cardinality of the set $\mathrm{Act}(10^{-3}, p)$ are

$\dim^{\mathrm{sprp}}$	$\lvert\mathrm{Act}\rvert$	β
4	109	2
3	22	3
2	10	4

for $p = 1$ and

$\dim^{\mathrm{sprp}}$	$\lvert\mathrm{Act}\rvert$	β
4	255	2
3	24	3
2	10	4

for $p = 2$.

Note that even for $\beta = 2$ and $p = 2$, we only need to approximate $255-1 = 254$ integrals[3] each with at most $\dim^{\mathrm{sprp}}(10^{-3}) = 4$ variables. Clearly, these integrals can be computed in parallel.

Remark 6.11. The method for constructing the smallest ε-active set $\mathrm{Act}(\varepsilon, p)$ from [8] is based on the fact that for product weights with $\gamma_j = c/j^{\beta}$, it is possible to approximate $\sum_{\mathfrak{u}\in\mathcal{U}}(\gamma_{\mathfrak{u}}\,\|S_{\mathfrak{u}}\|)^{p^*}$ with sufficiently high (given ε) accuracy, and next subtracting from the sum the largest values of $(\gamma_{\mathfrak{u}}\,\|S_{\mathfrak{u}}\|)^{p^*}$ until the remaining value is $\le \varepsilon^{p^*}$. Then the sequences/subsets $\mathfrak{u}$ corresponding to the subtracted terms are the elements of $\mathrm{Act}(\varepsilon, p)$. This procedure can be done (perhaps at large expense) for any weights in the case of a finite number d of variables (see Remark 6.2).

6.5 Multivariate decomposition method: anchored case

We assume in this section that the decomposition (6.2) is anchored, since then one can get information about the terms $f_{\mathfrak{u}}$ (which are unknown) from information about $f = \sum_{\mathfrak{u}\in\mathcal{U}_+} f_{\mathfrak{u}}$.

For simplicity, we will discuss here only information consisting of samples of f. For more general linear information, we refer to, e. g., [31, 33, 39].

As in [25], we use the following cost model: Let

$$\$: \mathbb{N} \cup \{0\} \to \mathbb{N}$$

be an increasing function with $\lim_{k\to\infty} \$(k) = \infty$. We assume that one can evaluate f only at points $[\boldsymbol{x}; \mathfrak{u}]$, and at the cost given by $\$(|\mathfrak{u}|)$. Recall, that $[\boldsymbol{x}; \mathfrak{u}]$ has all coordinates $x_j = 0$ for $j \notin \mathfrak{u}$. This agrees with the discussion in Section 6.2.1, where $f(\boldsymbol{x}) = g(\sum_{j=1}^{\infty} x_j\,\xi_j)$, and only finite sums $\sum_{j\in\mathfrak{u}} x_j\,\xi_j$ can be evaluated at a finite time. This also suggests that \$ should be at least linear. Since our positive results hold even for exponential functions, we assume in this section that

$$\Omega(k) = \$(k) = \exp(O(k)) \quad \text{as} \quad k \to \infty.$$

To evaluate the terms $f_{\mathfrak{u}}$, we use the following result from [26]:

$$f_{\mathfrak{u}}(\boldsymbol{x}) = \sum_{\mathfrak{v}\subseteq\mathfrak{u}} (-1)^{|\mathfrak{u}|-|\mathfrak{v}|} f([\boldsymbol{x}; \mathfrak{v}]).$$

Hence, the cost of obtaining the sample $f_{\mathfrak{u}}(\boldsymbol{x})$ is bounded by

$$(\text{cost of } f_{\mathfrak{u}}(\boldsymbol{x})) \le \sum_{\mathfrak{v}\subseteq\mathfrak{u}} \$(|\mathfrak{v}|) \le 2^{|\mathfrak{u}|}\,\$(|\mathfrak{u}|),$$

3 The empty set belongs to $\mathrm{Act}(10^{-3}, 2)$, but the integral of $f_{\emptyset}$ is just $f(\mathbf{0})$.

and

$$\text{if} \quad |\mathrm{u}| = O\left(\frac{\ln(1/\varepsilon)}{\ln(\ln(1/\varepsilon))}\right), \quad \text{then} \quad (\text{cost of} f_{\mathrm{u}}(\boldsymbol{x})) \leq \varepsilon^{-c/\ln(\ln(1/\varepsilon))}$$

for a positive constant c.

We assume that the weights are of product order-dependent form,

$$\gamma_{\mathrm{u}} = (|\mathrm{u}|!)^{\alpha} \prod_{j \in \mathrm{u}} \frac{c}{j^{\beta}} \quad \text{with} \quad \beta > \max(1, \alpha).$$

We also assume that F_{u} and S_{u} are $|\mathrm{u}|$-fold tensor products of F_1 and S_1, respectively, where F_1 is a space of univariate functions, and S_1 is the restriction of $\mathcal{S}$ to F_1. Of course, F_{u} and S_{u} are isomorphic to F_1 and S_1 if $|\mathrm{u}| = 1$. We assume that there are algorithms $A_{1,n}$ for approximating S_1 using n function values, and whose worst-case errors with respect to F_1 are bounded by

$$\text{error}(A_{1,n}; S_1, F_1) \leq \frac{c_1}{(n+1)^r} \tag{6.7}$$

for some $c_1, r > 0$.

The multivariate decomposition method is as follows: For given error demand $\varepsilon > 0$, construct an ε-active set $\mathcal{V}_\varepsilon = \mathcal{V}_{\varepsilon,1}$ for $p = 1$ or $\mathcal{V}_{\varepsilon,p,a}$ for $p > 1$, as in Theorem 6.8. Recall that $\|\mathcal{V}_\varepsilon\|_\infty = O(\ln(1/\varepsilon)/\ln(\ln(1/\varepsilon)))$. Next, for each $\mathrm{u} \in \mathcal{V}_\varepsilon$, consider algorithms $A_{\mathrm{u},n}$ for approximating $S_{\mathrm{u}}(f_{\mathrm{u}})$ obtained through Smolyak's construction applied to algorithms $A_{1,n}$, see e. g., [6, 30, 36]. Let $\text{error}(A_{\mathrm{u},n}; S_{\mathrm{u}}, F_{\mathrm{u}})$ denote the worst-case errors of $A_{\mathrm{u},n}$ for approximating S_{u}. Then one can choose n_{u} for all $\mathrm{u} \in \mathcal{V}_\varepsilon$ (see e. g. [33] for details) in such a way that

$$\sum_{\mathrm{u} \in \mathcal{V}_\varepsilon} n_{\mathrm{u}} \, \$(|\mathrm{u}|) = O(\varepsilon^{-\delta - \max(1/r, 1/(\beta - 1/p^*))}),$$

and

$$\left[\sum_{\mathrm{u} \in \mathcal{V}_\varepsilon} (\gamma_{\mathrm{u}} \, \text{error}(A_{\mathrm{u},n_{\mathrm{u}}}; S_{\mathrm{u}}, F_{\mathrm{u}}))^{p^*}\right]^{1/p^*} \leq \varepsilon.$$

Then the corresponding MDM, given by

$$\mathcal{A}_\varepsilon(f) := \sum_{\mathrm{u} \in \mathcal{V}_\varepsilon} A_{\mathrm{u},n_{\mathrm{u}}}(f_{\mathrm{u}}),$$

has the properties summarized in the following theorem:

Theorem 6.12. *Under the assumptions of this section, the worst-case error and the information cost of $\mathcal{A}_\varepsilon$ are, respectively, bounded by*

$$\text{error}(\mathcal{A}_\varepsilon; \mathcal{S}, \mathcal{F}_{\gamma,p}) \leq 2^{1/p^*} \varepsilon$$

and

$$\text{cost}(\mathcal{A}_\varepsilon) \leq c_\delta \, \varepsilon^{-\delta - \max(1/r, 1/(\beta - 1/p^*))} \quad \textit{for all } \delta > 0.$$

We end this section with the following remark.

Remark 6.13. If the exponent r in (6.7) is the largest possible among all algorithms for approximating S_1, then (modulo δ) the exponent $\max(1/r, 1/(\beta - 1/p^*))$ is optimal, which follows from a complexity lower bound in [25].

6.6 Other decompositions

For a large class of spaces $\mathcal{F}_{\gamma,p}$ with product weights γ, the norm defined via anchored decomposition is equivalent to the norms with other decompositions. If this is the case, then positive results for anchored spaces (e. g., small ε-truncation/superposition dimensions, or effectiveness of MDM's) are also applicable to not necessarily anchored spaces.

As mentioned in the introduction, the research on the equivalence of norms for different decompositions was initiated in [13].

The presentation in this section is based on results in [17]. As before, $\mathcal{F}_{\gamma,p}$ consists of $f = \sum_{\mathfrak{u}\in\mathcal{U}_+} f_{\mathfrak{u}}$ with $f_{\mathfrak{u}} \in F_{\mathfrak{u}}$, where now the spaces $F_{\mathfrak{u}}$ have the following form: For given positive and measurable function $\psi : D \to \mathbb{R}_+$, $q \in [1,\infty]$, and $\mathfrak{u} \neq \emptyset$, let

$$L_{\mathfrak{u},\psi,q} = \{g_{\mathfrak{u}} : D^{|\mathfrak{u}|} \to \mathbb{R} : \|g_{\mathfrak{u}}\|_{L_{\mathfrak{u},\psi,q}} < \infty\},$$

where

$$\|g_{\mathfrak{u}}\|_{L_{\mathfrak{u},\psi,q}} = \left[\int_{D^{|\mathfrak{u}|}} |g_{\mathfrak{u}}(\boldsymbol{t}_{\mathfrak{u}})|^q \, \psi_{\mathfrak{u}}(\boldsymbol{t}_{\mathfrak{u}}) \, \mathrm{d}\boldsymbol{t}_{\mathfrak{u}}\right]^{1/q} \quad \text{and} \quad \psi_{\mathfrak{u}}(\boldsymbol{x}_{\mathfrak{u}}) = \prod_{j\in\mathfrak{u}} \psi(x_j).$$

For a measurable function

$$\kappa : D \times D \to \mathbb{R} \quad \text{and} \quad \kappa_{\mathfrak{u}}(\boldsymbol{x}_{\mathfrak{u}}, \boldsymbol{t}_{\mathfrak{u}}) = \prod_{j\in\mathfrak{u}} \kappa(x_j, y_j),$$

the spaces $F_{\mathfrak{u}}$ are given by

$$F_{\mathfrak{u}} = K_{\mathfrak{u}}(L_{\mathfrak{u},\psi,q}) \quad \text{with} \quad f_{\mathfrak{u}} = K_{\mathfrak{u}}(g_{\mathfrak{u}}) = \int_{D^{|\mathfrak{u}|}} g_{\mathfrak{u}}(\boldsymbol{t}_{\mathfrak{u}}) \, \kappa_{\mathfrak{u}}(\cdot_{\mathfrak{u}}, \boldsymbol{t}_{\mathfrak{u}}) \, \mathrm{d}\boldsymbol{t}_{\mathfrak{u}}$$

endowed with the norm

$$\|f_{\mathfrak{u}}\|_{F_{\mathfrak{u}}} = \|K_{\mathfrak{u}}(g_{\mathfrak{u}})\|_{L_{\mathfrak{u},\psi,q}}.$$

We also assume that

$$\kappa(0, \cdot) \equiv 0,$$

so that the corresponding space $\mathcal{F}_{\gamma,p}$ and the decomposition $f = \sum_{\mathfrak{u}\in\mathcal{U}_+} f_{\mathfrak{u}}$ are anchored at zero.

Example 6.14. Clearly, the spaces $\mathcal{F}_{\gamma,p}$ defined as above, are special cases of spaces discussed in previous sections. For instance, $D = [0,1]$ and $\kappa(x,t) = \max(0, x-t)$, $\psi \equiv 1$ and $p = q$, yield the space from *Integration example.* $\kappa(x,t) = \max(0, (x-t)^{r-1}/(r-1)!)$ yields the space of functions with mixed partial derivatives of order r belonging to $L_q([0,1]^{|\mathrm{u}|})$.

For $D = [1,\infty)$, $\kappa(x,t) = \max(0, x-t)$, f_{u} are such that their mixed partial derivatives of order 1 belong to ψ_{u}-weighted $L_q(\mathbb{R}_+^{|\mathrm{u}|})$. Note that by a specific choice of ψ, one can make the spaces large or small. For instance, for $\psi(t) = \exp(-t)$, the corresponding derivatives may diverge very fast for $t \to \infty$. Moreover, for $q = \infty$, it is enough for these derivatives to be bounded by $c \cdot \exp(t)$.

As in Remark 5 of [17], consider a linear functional ℓ on the space $F_{\{1\}}$ of univariate functions such that

$$\ell\left(\int g(t)\,\kappa(\cdot,t)\,\mathrm{d}t\right) = \int_D g(t)\,\ell(\kappa(\cdot,t))\,\mathrm{d}t \quad \text{for all } g \in L_{\{1\},q,\psi}(D).$$

We also assume that ℓ is continuous, i. e.,

$$\|\bar{\ell}_q\|_{L_{q^*}} < \infty \quad \text{for} \quad \bar{\ell}_q(t) = \frac{\ell(\kappa(\cdot,t))}{\psi^{1/q}(t)}.$$

Let

$$K_{\mathrm{u}}^{[\ell]}(g_{\mathrm{u}})(\boldsymbol{x}_{\mathrm{u}}) := \int_{D^{|\mathrm{u}|}} g_{\mathrm{u}}(\boldsymbol{t}_{\mathrm{u}}) \prod_{j\in\mathrm{u}} (\kappa(x_j,t_j) - \ell(\kappa(\cdot_j,t_j)))\,\mathrm{d}\boldsymbol{t}_{\mathrm{u}},$$

$$F_{\mathrm{u}}^{[\ell]} := K_{\mathrm{u}}^{[\ell]}(L_{\mathrm{u},\psi,q}),$$

and $\mathcal{F}_{\gamma,p}^{[\ell]}$ contains $\sum_{\mathrm{u}\in\mathcal{U}_+} f_{\mathrm{u},\ell}$ with $f_{\mathrm{u},\ell} = K_{\mathrm{u},\ell}(g_{\mathrm{u}})$, with finite norm

$$\|f\|_{\mathcal{F}_{\gamma,p}^{[\ell]}} = \left[\sum_{\mathrm{u}\in\mathcal{U}_+} \gamma_{\mathrm{u}}^{-p}\,\|f_{\mathrm{u}}\|_{F_{\mathrm{u},q}^{[\ell]}}^p\right]^{1/p}.$$

Note that the decomposition is, in general, not anchored, and instead we have

$$\ell_j(f_{\mathrm{u},\ell}) = 0 \quad \text{if } j \in \mathrm{u},$$

where ℓ_j is the functional ℓ acting on jth variable, i. e.,

$$\ell_j = \bigotimes_{n=1}^{\infty} T_n \quad \text{with} \quad T_n = \begin{cases} \mathrm{id} & \text{if } n \neq j, \\ \ell & \text{if } n = j. \end{cases}$$

Integration example (continued). For functional $\ell(f) = \int_0^1 f(t)\,\mathrm{d}t$, the corresponding decomposition $f = \sum_{\mathrm{u}\in\mathcal{U}_+} f_{\mathrm{u},\ell}$ is the famous *ANOVA* decomposition.

The following Theorem is from [17]:

Theorem 6.15. *Assume that the weights have product form* $\gamma_{\mathfrak{u}} = \prod_{j\in\mathfrak{u}} \gamma_j$ *with finite* $\sum_{j=1}^{\infty} \gamma_j$. *Assume also that* $\|\bar{\ell}_q\|_{L_{q^*}} < \infty$ *for all* $q \in [1,\infty]$. *Then the spaces* $\mathcal{F}_{\gamma,p}$ *and* $\mathcal{F}_{\gamma,p}^{[\ell]}$ *are equal (as the sets of functions). Moreover, the embedding*

$$\imath : \mathcal{F}_{\gamma,p} \to \mathcal{F}_{\gamma,p}^{[\ell]}$$

satisfies

$$\|\imath\| = \|\imath^{-1}\| < \infty.$$

For sharp upper and lower bounds on $\|\imath\|$ for all $p, q \in [1,\infty]$, we refer to Proposition 8 and Corollary 9 in [17]. Since the spaces (as sets) are equal, the functions f_k from the definition of truncation dimension also belong to $\mathcal{F}_{\gamma,p}^{[\ell]}$. This means that we can extend the definition of ε-truncation and superposition dimensions to the spaces $\mathcal{F}_{\gamma,p}^{[\ell]}$. Moreover, the algorithm $\mathcal{A}_\varepsilon$ is also well defined for $\mathcal{F}_{\gamma,p}^{[\ell]}$. We have the following corollary, which states that both ε dimensions and the worst-case error of $\mathcal{A}_\varepsilon$ are small also for the norm $\|\cdot\|_{\mathcal{F}_{\gamma,p}^{[\ell]}}$.

Corollary 6.16. *Let* $\dim^\star(\varepsilon, \mathcal{S}; \mathcal{F}_{\gamma,p}^{[\ell]})$ *for* $\star \in \{\mathrm{trnc}, \mathrm{sprp}\}$ *denote the corresponding* ε*-truncation and* ε*-superposition dimensions with respect to the norm of* $\mathcal{F}_{\gamma,p}^{[\ell]}$. *Then under the assumptions of Theorem* 6.15, *we have*

$$\dim^\star(\varepsilon\,\|\imath\|, \mathcal{S}) \le \dim^\star(\varepsilon, \mathcal{S}; \mathcal{F}_{\gamma,p}^{[\ell]}) \le \dim^\star(\varepsilon/\|\imath\|, \mathcal{S}).$$

Moreover, under the assumptions of Theorem 6.12 *and for product weights, the algorithm* $\mathcal{A}_\varepsilon$ *has the worst-case error with respect to the space* $\mathcal{F}_{\gamma,p}^{[\ell]}$ *bounded by*

$$\mathrm{error}(\mathcal{A}_\varepsilon; \mathcal{S}, \mathcal{F}_{\gamma,p}^{[\ell]}) \le 2^{1/p^*}\,\|\imath\|\,\varepsilon.$$

Bibliography

[1] J. Baldeaux and M. Gnewuch. Optimal randomized multilevel algorithms for infinite-dimensional integration of function spaces with ANOVA-type decomposition, *SIAM J. Numer. Anal.* **52** (2014), 1128–1155.

[2] J. Creutzig, S. Dereich, T. Müller-Gronbach, and K. Ritter. Infinite-dimensional quadrature and approximation of distributions, *Fund. Comput. Math.* **9** (2009), 391–429.

[3] J. Dick and M. Gnewuch. Infinite-dimensional integration in weighted Hilbert spaces: anchored decompositions optimal deterministic algorithms, and higher order convergence, *Found. Comput. Math.* **14** (2014), 1027–1077.

[4] J. Dick and M. Gnewuch. Optimal randomized changing dimension algorithms for infinite-dimensional integration on function spaces with ANOVA-type decomposition, *J. Approx. Theory*, **184** (2014), 111–145.

[5] D. Dũng and M. Griebel. Hyperbolic cross approximation in infinite dimensions, *J. Complex.* **33** (2016), 55–88.

[6] T. Gernstner and M. Griebel. Numerical integration using sparse grids, *Numer. Algorithms* **18** (1998), 209–232.

[7] A. Gilbert, F. Y. Kuo, D. Nuyens, and G. W. Wasilkowski. Efficient implementations of the Multivariate Decomposition Method for approximating infinite-variate integrals, *SIAM J. Sci. Comput.* **40-5** (2018), A3240–A3266.

[8] A. Gilbert and G. W. Wasilkowski. Small superposition dimension and active set construction for multivariate integration under modest error demand, *J. Complex.* **42** (2017), 94–109.

[9] M. Gnewuch, M. Hefter, A. Hinrichs, and K. Ritter. Embeddings of weighted Hilbert spaces and applications to multivariate and infinite-dimensional integration, *J. Approx. Theory* **222** (2017), 8–39.

[10] M. Gnewuch, M. Hefter, A. Hinrichs, K. Ritter, and G. W. Wasilkowski. Equivalence of weighted anchored and ANOVA spaces of functions with mixed smoothness of order one in L_p, *J. Complex.* **40** (2017), 78–99.

[11] M. Gnewuch, M. Hefter, A. Hinrichs, K. Ritter, and G. W. Wasilkowski. Embeddings for infinite-dimensional integration and L_2-approximation with increasing smoothness, *J. Complex.* **54** (2019), 1–32.

[12] M. Gnewuch, S. Mayer, and K. Ritter. On weighted Hilbert spaces and integration of functions of infinitely many variables, *J. Complex.* **30** (2014), 29–47.

[13] M. Hefter and K. Ritter. On embeddings of weighted tensor product Hilbert spaces, *J. Complex.* **31** (2015), 405–423.

[14] M. Hefter, K. Ritter, and G. W. Wasilkowski. On equivalence of weighted anchored and ANOVA spaces of functions with mixed smoothness of order one in L_1 or L_∞, *J. Complex.* **32** (2016), 1–19.

[15] F. J. Hickernell, T. Müller-Gronbach, B. Niu, and K. Ritter. Multi-level Monte Carlo algorithms for infinite-dimensional integration on $\mathbb{R}^{\mathbb{N}}$, *J. Complex.* **26** (2010), 229–254.

[16] F. J. Hickernell and X. Wang. The error bounds and tractability of quasi-Monte Carlo algorithms in infinite dimension, *Math. Comput.* **71** (2001), 1641–1661.

[17] A. Hinrichs, P. Kritzer, F. Pillichshammer, and G. W. Wasilkowski. Truncation dimension for linear problems on multivariate function spaces, *Numer. Algorithms* **80** (2019), 661–685.

[18] A. Hinrichs and J. Schneider. Equivalence of anchored and ANOVA spaces via interpolation, *J. Complex.* **33** (2016), 190–198.

[19] P. Kritzer, F. Pillichshammer, and G. W. Wasilkowski. Very low truncation dimension for high dimensional integration under modest error demand, *J. Complex.* **35** (2016), 63–85.

[20] P. Kritzer, F. Pillichshammer, and G. W. Wasilkowski. A note on equivalence of anchored and ANOVA spaces; lower bounds, *J. Complex.* **38** (2017), 31–38.

[21] P. Kritzer, F. Pillichshammer, and G. W. Wasilkowski. Truncation in average and worst case settings for special classes of ∞-variate functions, *Math. Comput. Simul.* **161** (2019), 52–65.

[22] F. Y. Kuo and D. Nuyens. Application of quasi-Monte Carlo methods to elliptic PDE's with random diffusion coefficients – a survey of analysis and implementation, *Found. Comput. Math.* **16** (2016), 1631–1696.

[23] F. Y. Kuo, D. Nuyens, L. Plaskota, I. H. Sloan, and G. W. Wasilkowski. Infinite-dimensional integration and the multivariate decomposition method, *J. Comput. Appl. Math.* **326** (2017), 217–234.

[24] F. Y. Kuo, Ch. Schwab, and I. H. Sloan. Quasi-Monte Carlo finite element methods for a class of elliptic partial differential equations with random coefficients, *SIAM J. Numer. Anal.* **50** (2012), 3351–3374.

[25] F. Y. Kuo, I. H. Sloan, G. W. Wasilkowski, and H. Woźniakowski. Liberating the dimension, *J. Complex.* **26** (2010), 422–454.
[26] F. Y. Kuo, I. H. Sloan, G. W. Wasilkowski, and H. Woźniakowski. On decompositions of multivariate functions, *Math. Comput.* **79** (2010), 953–966.
[27] L. Plaskota and G. W. Wasilkowski. Tractability of infinite-dimensional integration in the worst case and randomized settings, *J. Complex.* **27** (2011), 505–518.
[28] L. Plaskota, G. W. Wasilkowski, and H. Woźniakowski. A new algorithm and worst case complexity for Feynman-Kac path integration, *J. Comput. Phys.* **164** (2000), 335–353.
[29] I. H. Sloan and H. Woźniakowski. When are quasi-Monte Carlo algorithms efficient for high-dimensional integrals? *J. Complex.* **14** (1998), 1–33.
[30] S. A. Smolyak. Quadrature and interpolation formulas for tensor products of certain classes of functions, *Dokl. Akad. Nauk SSSR* **4** (1963), 240–243.
[31] G. W. Wasilkowski. Liberating the dimension for L_2-approximation, *J. Complex.* **28** (2012), 304–319.
[32] G. W. Wasilkowski. Liberating the dimension for function approximation and integration, *Monte Carlo and Quasi-Monte Carlo Methods 2010* (L. Plaskota and H. Woźniakowski, eds.), pp. 211–231, Springer Proceedings in Math. and Stat., Springer-Verlag, Berlin, 2012.
[33] G. W. Wasilkowski. On tractability of linear tensor product problems for ∞-variate classes of functions, *J. Complex.* **29** (2013), 351–369.
[34] G. W. Wasilkowski. Tractability of approximation of ∞-variate functions with bounded mixed partial derivatives, *J. Complex.* **30** (2014), 325–346.
[35] G. W. Wasilkowski. Average case tractability of approximating ∞-variate functions, *Math. Comput.*, **83** (2014), 1319–1336.
[36] G. W. Wasilkowski and H. Woźniakowski. Explicit cost bounds of algorithms for multivariate tensor product problems, *J. Complex.* **11** (1995), 1–56.
[37] G. W. Wasilkowski and H. Woźniakowski. On tractability of path integration, *J. Math. Phys.* **37** (1996) 2071–2088.
[38] G. W. Wasilkowski and H. Woźniakowski. On the complexity of stochastic integration, *Math. Comput.* **70** (2000), 685–698.
[39] G. W. Wasilkowski and H. Woźniakowski. Liberating the dimension for function approximation, *J. Complex.* **27** (2011), 86–110.
[40] G. W. Wasilkowski and H. Woźniakowski. Liberating the dimension for function approximation: standard information, *J. Complex.* **27** (2011), 417–440.

Yuhan Ding, Fred J. Hickernell, Peter Kritzer, and Simon Mak

7 Adaptive approximation for multivariate linear problems with inputs lying in a cone

Abstract: We study adaptive approximation algorithms for general multivariate linear problems, where the sets of input functions are nonconvex cones. Whereas it is known that adaptive algorithms perform essentially no better than nonadaptive algorithms for convex and symmetric input sets, the situation may be different for nonconvex sets. The setting considered here is function approximation based on series expansions. Given an error tolerance, we use series coefficients of the input to construct an approximate solution such that the error does not exceed this tolerance. We study the situation, where we can bound the norm of the input based on a pilot sample, and the situation, where we keep track of the decay rate of the series coefficients of the input. Moreover, we consider situations, where it makes sense to infer coordinate and smoothness importance. Besides performing an error analysis, we also derive upper bounds on the information cost of our algorithms and lower bounds on the computational complexity of our problems, and we identify conditions, under which we can avoid a curse of dimensionality.

Keywords: Adaptive algorithms, approximation of linear problems, computational complexity, cones, series spaces, tractability

MSC 2010: 65D15, 68Q25, 41A25, 41A63

Acknowledgement: The authors would like to thank two anonymous referees for their suggestions, which helped to improve the presentation of the results. Furthermore, the authors thank Stefan Heinrich and Aicke Hinrichs for helpful comments. F. J. Hickernell, P. Kritzer, and S. Mak gratefully acknowledge the support of the Statistical and Applied Mathematical Sciences Institute year-long "Program on Quasi-Monte Carlo and High-Dimensional Sampling Methods for Applied Mathematics" through NSF-DMS-1638521. F. J. Hickernell also acknowledges the support of NSF-DMS-1522687. F. J. Hickernell and P. Kritzer thank the RICAM Special Semester Program 2018 for support. P. Kritzer gratefully acknowledges support by the Austrian Science Fund (FWF) Project F5506-N26, which is part of the Special Research Program "Quasi-Monte Carlo Methods: Theory and Applications".

Yuhan Ding, Fred J. Hickernell, Department of Applied Mathematics, Illinois Institute of Technology, RE 220, 10 W. 32nd Street, Chicago, IL 60616, USA, e-mails: yding2@hawk.iit.edu, hickernell@iit.edu
Peter Kritzer, Johann Radon Institute for Computational and Applied Mathematics (RICAM), Austrian Academy of Sciences, Altenbergerstr. 69, 4040 Linz, Austria, e-mail: peter.kritzer@oeaw.ac.at
Simon Mak, Department of Statistical Science, Duke University, Durham, NC 27708-0251, USA, e-mail: sm769@duke.edu

https://doi.org/10.1515/9783110635461-007

7.1 Introduction

In many situations, adaptive algorithms can be rigorously shown to perform *essentially no better* than nonadaptive algorithms. Yet, in practice adaptive algorithms are appreciated, because they relieve the user from stipulating the computational effort required to achieve the desired accuracy. The key to resolving this seeming contradiction is to construct a theory based on assumptions that favor adaptive algorithms. We do that here.

Adaptive algorithms infer the necessary computational effort based on the function data sampled. Adaptive algorithms may perform better than nonadaptive algorithms if the set of input functions is *nonconvex* or *nonsymmetric*. Here, we focus on the former situation and construct adaptive algorithms for general multivariate linear problems, where the input functions lie in nonconvex cones, $\mathcal{C}$. By a cone, we mean a set of functions, where constant multiples of every function in the set also lie in the set. The precise definitions of the cones that we study are given below.

Our algorithms use a finite number of series coefficients of the input function to construct an approximate solution that satisfies an absolute error tolerance. We show our algorithms to be essentially optimal. We derive conditions under which the problem is tractable, i. e., the information cost of constructing the approximate solution does not increase exponentially with the dimension of the input function domain. In the remainder of this section we outline the general idea of this chapter.

We study the approximation of a linear solution operator $\mathrm{SOL} : \mathcal{F} \to \mathcal{G}$, which maps $f \in \mathcal{F}$ to some $g \in \mathcal{G}$, where we assume that $\mathcal{F}$ and $\mathcal{G}$ are Banach spaces with norms $\|\cdot\|_{\mathcal{F}}$ and $\|\cdot\|_{\mathcal{G}}$, respectively. Our goal is to define a suitable set of input functions $\mathcal{C}$, and an algorithm $\mathrm{ALG}(\cdot,\cdot) : \mathcal{C} \times (0,\infty) \to \mathcal{G}$ that satisfies the error criterion:

$$\|\mathrm{SOL}(f) - \mathrm{ALG}(f,\varepsilon)\|_{\mathcal{G}} \le \varepsilon \quad \forall \varepsilon > 0,\ f \in \mathcal{C}, \tag{7.1}$$

where ε denotes the error tolerance. Such an algorithm will contain a rule for choosing the number of series coefficients required, $n(f,\varepsilon)$, which depends on f and ε. The objectives of this chapter are to

- construct such a rule,
- choose a set $\mathcal{C}$ of input functions, for which the rule is valid,
- characterize the information cost of ALG,
- determine whether ALG has optimal information cost, and
- understand the dependence of this cost on the number of input variables, d, as well as the error tolerance, ε.

We assume in this chapter that the functions $f \in \mathcal{F}$ can be represented in terms of a basis representation as $f = \sum_{\boldsymbol{k}\in\mathbb{K}} \widehat{f}(\boldsymbol{k}) u_{\boldsymbol{k}}$, and we allow algorithms $\mathrm{ALG}(f,\varepsilon)$ to use $n(f,\varepsilon)$ coefficients $\widehat{f}(\boldsymbol{k})$ of this representation. That is, we study algorithms that have an output of the form

$$\mathrm{ALG}(f,\varepsilon) = \sum_{i=1}^{n(f,\varepsilon)} \widehat{f}(\boldsymbol{k}_i)v_{\boldsymbol{k}_i} \quad \forall f \in \mathcal{F},\ n(f,\varepsilon) \in \mathbb{N},$$

where an oracle provides $\widehat{f}(\boldsymbol{k}_1),\dots,\widehat{f}(\boldsymbol{k}_{n(f,\varepsilon)})$ with $n(f,\varepsilon)$ being chosen based on a stopping rule. We assume that it is possible to sample the series coefficients of the input function. This is a more restrictive assumption than being able to sample any linear functional, but it is different than only being able to sample function values. An important future problem is to extend the theory in this chapter to the case, where the only function data available are function values.

The information cost of $\mathrm{ALG}(f,\varepsilon)$ is denoted $\mathrm{COST}(\mathrm{ALG},f,\varepsilon)$ and defined as the number of function data—in our situation, series coefficients—required by $\mathrm{ALG}(f,\varepsilon)$, i. e., $\mathrm{COST}(\mathrm{ALG},f,\varepsilon) = n(f,\varepsilon)$. For adaptive algorithms this cost varies with the input function f. We call an algorithm *nonadaptive* if $n(f,\varepsilon) = n(\varepsilon)$, independently of the particular choice of $f \in \mathcal{C}$, and also the sequence $\boldsymbol{k}_1,\boldsymbol{k}_2,\dots$ is independent of the choice of $f \in \mathcal{C}$.

In the following, let $\mathcal{B}_R := \{f \in \mathcal{F} : \|f\|_{\mathcal{F}} \le R\}$ denote the ball of radius R in the space of input functions. We now define the information cost of the algorithm in general, recognizing that it will tend to depend on $\|f\|_{\mathcal{F}}$:

$$\mathrm{COST}(\mathrm{ALG},\mathcal{C},\varepsilon,R) := \max_{f\in\mathcal{C}\cap\mathcal{B}_R} \mathrm{COST}(\mathrm{ALG},f,\varepsilon).$$

Note that whereas the cost depends on $\|f\|_{\mathcal{F}}$, $\mathrm{ALG}(f,\varepsilon)$ has no knowledge of f beyond the fact that it lies in $\mathcal{C}$. It is common for $\mathrm{COST}(\mathrm{ALG},\mathcal{C},\varepsilon,R)$ to be $\mathcal{O}(\varepsilon^{-p})$, or perhaps asymptotically $c\log(1+\varepsilon^{-1})$.

A crucial concept is the set of all possible algorithms that may be constructed using series coefficients and that *satisfy error criterion* (7.1). We denote this set by $\mathcal{A}(\mathcal{C})$. Furthermore, we define the *computational complexity* of a problem as the information cost of the best algorithm:

$$\mathrm{COMP}(\mathcal{C},\varepsilon,R) := \min_{\mathrm{ALG}\in\mathcal{A}(\mathcal{C})} \mathrm{COST}(\mathrm{ALG},\mathcal{C},\varepsilon,R).$$

These definitions follow the information-based complexity literature [12, 13]. We define an algorithm to be *essentially optimal* if there exist some fixed positive ω, $\varepsilon_{\max}$, and $R_{\min}$ for which

$$\mathrm{COST}(\mathrm{ALG},\mathcal{C},\varepsilon,R) \le \mathrm{COMP}(\mathcal{C},\omega\varepsilon,R) \quad \forall \varepsilon \in (0,\varepsilon_{\max}],\ R \in [R_{\min},\infty). \tag{7.2}$$

If the complexity of the problem is $\mathcal{O}(\varepsilon^{-p})$, the cost of an essentially optimal algorithm is also $\mathcal{O}(\varepsilon^{-p})$. If the complexity of the problem is asymptotically $c\log(1+\varepsilon^{-1})$, then the cost of an essentially optimal algorithm is also asymptotically $c\log(1+\varepsilon^{-1})$. We will show that our adaptive algorithms presented in Sections 7.3 and 7.4 are essentially optimal.

We demonstrate below that nonadaptive algorithms are optimal when the set of inputs is chosen to be $\mathcal{C} = \mathcal{B}_R$. However, typically one cannot bound the norm of the input function a priori, so nonadaptive algorithms may be impractical. This will lead to *adaptive* algorithms ALG satisfying error criterion (7.1). For such algorithms, the set of allowable input functions, $\mathcal{C}$, will be a *cone*, not a ball. We study different variants of cones in different sections of this chapter.

The first definition of a cone is given in (7.18), where the idea is that the cone consists of $f \in \mathcal{F}$ such that

$$\|f\|_{\mathcal{F}} \leq h(\hat{f}(\boldsymbol{k}_1), \dots, \hat{f}(\boldsymbol{k}_{n_1}))$$

for a suitably chosen homogeneous function $h : \mathbb{R}^{n_1} \to [0, \infty)$. This means that the norm of f can be bounded well in terms of a *pilot sample* of size $n_1 \in \mathbb{N}$. See (7.18) for the precise definition. In Section 7.4.1, we will modify the definition of cones such that we can keep track of the decay rate of the series coefficients of the input function (see (7.32)), and in Section 7.5 we allow a situation, where we infer coordinate and smoothness importance from some initial data, thus making the definition of the cone more practical (see (7.46)).

Note that algorithms satisfying error criterion (7.1) cannot exist for $\mathcal{C} = \mathcal{F}$. Any algorithm must require a finite sample size, even if it is huge. Then, there must exist some $f \in \mathcal{F}$ that look exactly like the zero function to the algorithm, but for which $\|\mathrm{SOL}(f)\|_{\mathcal{G}}$ is arbitrarily large. Thus, algorithms satisfying the error criterion exist only for some strict subset of $\mathcal{F}$, such as a ball or a cone. Identifying that subset is both an art and a science.

The rest of this chapter is structured as follows: In the next section, we formulate our problem more formally, including the precise definition of the input space, $\mathcal{F}$, in terms of a sequence of weights, $\boldsymbol{\lambda}$, which reflect the assumed smoothness of the input functions. We point out some results for nonadaptive algorithms with inputs from balls. Furthermore, we illustrate our theory by a particular example, and mention some definitions regarding *tractability*.

In Section 7.3, we define a cone of input functions, $\mathcal{C}$, in (7.18) whose norms can be bounded above in terms of the series coefficients obtained from a pilot sample. Adaptive Algorithm 7.2 is shown to be optimal for this $\mathcal{C}$. We also identify necessary and sufficient conditions for tractability.

Section 7.4 considers the situation, where function data is relatively inexpensive, and we track the decay rate of the series coefficients. Adaptive Algorithm 7.3 is shown to be optimal in this situation.

Section 7.5 considers the case where the most suitable weights $\boldsymbol{\lambda}$, which are used to define the input space, $\mathcal{F}$, are not known a priori and are instead themselves inferred from function data. Adaptive Algorithm 7.4 combines this inference step with Algorithm 7.2 to construct an approximation that satisfies the error criterion.

7.2 Formal setting

7.2.1 Problem formulation

Now, we define our problem more formally. The solution operator SOL maps the input function to an output, SOL : $\mathcal{F} \to \mathcal{G}$. The Banach spaces of inputs and outputs are defined by series expansions:

$$\mathcal{F} := \left\{ f = \sum_{\boldsymbol{k} \in \mathbb{K}} \widehat{f}(\boldsymbol{k}) u_{\boldsymbol{k}} : \|f\|_{\mathcal{F}} := \left\| \left(\frac{\widehat{f}(\boldsymbol{k})}{\lambda_{\boldsymbol{k}}} \right)_{\boldsymbol{k} \in \mathbb{K}} \right\|_{\rho} < \infty \right\}, \quad \text{with} \quad 1 \le \rho \le \infty,$$

and

$$\mathcal{G} := \left\{ g = \sum_{\boldsymbol{k} \in \mathbb{K}} \widehat{g}(\boldsymbol{k}) v_{\boldsymbol{k}} : \|g\|_{\mathcal{G}} := \left\| (\widehat{g}(\boldsymbol{k}))_{\boldsymbol{k} \in \mathbb{K}} \right\|_{\tau} < \infty \right\}, \quad \text{with} \quad 1 \le \tau \le \rho.$$

Here, $\{u_{\boldsymbol{k}}\}_{\boldsymbol{k} \in \mathbb{K}}$ is a basis for the input Banach space $\mathcal{F}$, $\{v_{\boldsymbol{k}}\}_{\boldsymbol{k} \in \mathbb{K}}$ is a basis for the output Banach space $\mathcal{G}$, and $\mathbb{K}$ is a countable index set. These bases are defined to match the solution operator:

$$\mathrm{SOL}(u_{\boldsymbol{k}}) = v_{\boldsymbol{k}} \quad \forall \boldsymbol{k} \in \mathbb{K}. \tag{7.3}$$

The $\lambda_{\boldsymbol{k}}$ represent the importance of the series coefficients of the input function. The larger $\lambda_{\boldsymbol{k}}$ is, the more important $\widehat{f}(\boldsymbol{k})$ is. The $\lambda_{\boldsymbol{k}}$ are assumed to have a known order:

$$\lambda_{\boldsymbol{k}_1} \ge \lambda_{\boldsymbol{k}_2} \ge \cdots > 0, \tag{7.4}$$

which implies an ordering of the wavenumbers, $\boldsymbol{k}$. We also assume that all $\lambda_{\boldsymbol{k}}$ are positive to avoid the trivial case, where SOL(f) can be expressed exactly as a finite sum for all $f \in \mathcal{F}$. We denote the sequence of weights $(\lambda_{\boldsymbol{k}_i})_{i=1}^{\infty}$ by $\boldsymbol{\lambda}$.

Although this problem formulation is quite general in some aspects, condition (7.3) is somewhat restrictive. In principle, the choice of basis can be made via the singular value decomposition, but in practice, if the norms of $\mathcal{F}$ and $\mathcal{G}$ are specified without reference to their respective bases, it may be difficult to identify bases satisfying (7.3).

Some of the results on balls in this chapter can be found elsewhere, e. g., [10, Section 11.11] and [11, Chap. VI, Sect. 2], but we provide proofs for the sake of completeness. To facilitate our derivations below, we establish the following lemma via Hölder's inequality:

Lemma 7.1. *Let $\mathcal{K}$ be some proper or improper subset of the index set $\mathbb{K}$. Moreover, let ρ' be defined by the relation*

$$\frac{1}{\rho} + \frac{1}{\rho'} = \frac{1}{\tau}, \quad i.e., \rho' := \frac{\rho\tau}{\rho - \tau},$$

so $\tau \le \rho' \le \infty$. *Let* $\Lambda := \|(\lambda_{\boldsymbol{k}})_{\boldsymbol{k}\in\mathcal{K}}\|_{\rho'}$ *be the norm of a subset of the weights. Then the following are true for* $f = \sum_{\boldsymbol{k}\in\mathcal{K}} \widehat{f}(\boldsymbol{k}) u_{\boldsymbol{k}}$*:*

$$\|\mathrm{SOL}(f)\|_{\mathcal{G}} = \|(\widehat{f}(\boldsymbol{k}))_{\boldsymbol{k}\in\mathcal{K}}\|_{\tau} \le \|f\|_{\mathcal{F}}\,\Lambda, \tag{7.5}$$

$$|\widehat{f}(\boldsymbol{k})| = \begin{cases} \dfrac{R\lambda_{\boldsymbol{k}}^{\rho'/\rho+1}}{\Lambda^{\rho'/\rho}}, & \forall \boldsymbol{k} \in \mathcal{K}, \quad \text{if } \rho' < \infty, \\ R\Lambda\delta_{\boldsymbol{k},\boldsymbol{k}^*}, & \forall \boldsymbol{k} \in \mathcal{K},\ \boldsymbol{k}^* \in \mathcal{K} \text{ satisfies } \lambda_{\boldsymbol{k}^*} = \Lambda, \quad \text{if } \rho' = \infty, \end{cases}$$

$$\implies \quad \|f\|_{\mathcal{F}} = R \text{ and } \|\mathrm{SOL}(f)\|_{\mathcal{G}} = R\Lambda. \tag{7.6}$$

Equality (7.6) *illustrates how inequality* (7.5) *may be made tight.*

Proof. We give the proof for $\rho' < \infty$. The proof for $\rho' = \infty$ follows similarly. The proof of inequality (7.5) proceeds by applying Hölder's inequality:

$$\begin{aligned} \|\mathrm{SOL}(f)\|_{\mathcal{G}} &= \Big\|\sum_{\boldsymbol{k}\in\mathcal{K}} \widehat{f}(\boldsymbol{k}) v_{\boldsymbol{k}}\Big\|_{\mathcal{G}} = \|(\widehat{f}(\boldsymbol{k}))_{\boldsymbol{k}\in\mathcal{K}}\|_{\tau} = \Bigg[\sum_{\boldsymbol{k}\in\mathcal{K}} \bigg|\frac{\widehat{f}(\boldsymbol{k})}{\lambda_{\boldsymbol{k}}}\bigg|^{\tau} \lambda_{\boldsymbol{k}}^{\tau}\Bigg]^{1/\tau} \\ &\le \bigg\|\bigg(\frac{\widehat{f}(\boldsymbol{k})}{\lambda_{\boldsymbol{k}}}\bigg)_{\boldsymbol{k}\in\mathcal{K}}\bigg\|_{\rho} \|(\lambda_{\boldsymbol{k}})_{\boldsymbol{k}\in\mathcal{K}}\|_{\rho'} = \|f\|_{\mathcal{F}}\,\Lambda \quad \text{since } \frac{1}{\rho} + \frac{1}{\rho'} = \frac{1}{\tau}. \end{aligned} \tag{7.7}$$

Substituting the formula for $|\widehat{f}(\boldsymbol{k})|$ in (7.6) into equation (7.7) and applying the relationship between ρ, ρ', and τ yields

$$\|\mathrm{SOL}(f)\|_{\mathcal{G}} = \frac{R\|(\lambda_{\boldsymbol{k}}^{\rho'/\rho+1})_{\boldsymbol{k}\in\mathcal{K}}\|_{\tau}}{\Lambda^{\rho'/\rho}} = \frac{R\|(\lambda_{\boldsymbol{k}})_{\boldsymbol{k}\in\mathcal{K}}\|_{\rho'}^{\rho'/\rho+1}}{\Lambda^{\rho'/\rho}} = R\Lambda.$$

Moreover,

$$\|f\|_{\mathcal{F}} = \bigg\|\bigg(\frac{\widehat{f}(\boldsymbol{k})}{\lambda_{\boldsymbol{k}}}\bigg)_{\boldsymbol{k}\in\mathcal{K}}\bigg\|_{\rho} = \frac{R\|(\lambda_{\boldsymbol{k}}^{\rho'/\rho})_{\boldsymbol{k}\in\mathcal{K}}\|_{\rho}}{\Lambda^{\rho'/\rho}} = \frac{R\|(\lambda_{\boldsymbol{k}})_{\boldsymbol{k}\in\mathcal{K}}\|_{\rho'}^{\rho'/\rho}}{\Lambda^{\rho'/\rho}} = R.$$

This completes the proof. □

Taking $\mathcal{K} = \mathbb{K}$ in the lemma above, the norm of the solution operator can be expressed in terms of the norm of $\boldsymbol{\lambda}$ as follows:

$$\|\mathrm{SOL}\|_{\mathcal{F}\to\mathcal{G}} = \sup_{\|f\|_{\mathcal{F}}\le 1} \|\mathrm{SOL}(f)\|_{\mathcal{G}} = \|\boldsymbol{\lambda}\|_{\rho'}. \tag{7.8}$$

We assume throughout this chapter that the weights are chosen to keep this norm finite, namely,

$$\|\boldsymbol{\lambda}\|_{\rho'} < \infty. \tag{7.9}$$

7.2.2 An illustrative example

Consider the case of approximating functions defined over $[-1,1]^d$, using a Chebyshev polynomial basis. The input function is denoted f, and the solution is $\mathrm{SOL}(f) = f$. In this case,

$$f = \sum_{\boldsymbol{k}\in\mathbb{N}_0^d} \widehat{f}(\boldsymbol{k}) u_{\boldsymbol{k}} =: \mathrm{SOL}(f), \quad \boldsymbol{k} = (k_1,\dots,k_d) \in \mathbb{N}_0^d,$$

$$u_{\boldsymbol{k}} := \prod_{\ell=1}^{d} u_{k_\ell}, \quad u_k(x) := \cos(k\cos^{-1}(x)) \quad \forall k \in \mathbb{N}_0.$$

For this example, let $\mathcal{F}$ denote a Hilbert space of input functions, where we choose $\mathbb{K} = \mathbb{N}_0^d$ and $\rho = 2$,

$$\|f\|_{\mathcal{F}} := \left\| \left(\frac{\widehat{f}(\boldsymbol{k})}{\lambda_{\boldsymbol{k}}} \right)_{\boldsymbol{k}\in\mathbb{N}_0^d} \right\|_2, \quad \text{where } \lambda_{\boldsymbol{k}} := \prod_{\substack{\ell=1 \\ k_\ell>0}}^{d} \frac{w_\ell}{k_\ell^r}, \quad r > 0. \tag{7.10}$$

The w_ℓ are nonnegative *coordinate weights*, which embody the assumption that f may depend more strongly on coordinates with larger w_ℓ than those with smaller w_ℓ. The definition of the $\mathcal{F}$-norm implies that an input function must have series coefficients that decay quickly enough as the degree of the polynomial increases. Larger r implies smoother input functions.

Define the norm of the Hilbert space of outputs as

$$\left\| \sum_{\boldsymbol{k}\in\mathbb{N}_0^d} \widehat{g}(\boldsymbol{k}) u_{\boldsymbol{k}} \right\|_{\mathcal{G}} := \|(\widehat{g}(\boldsymbol{k}))_{\boldsymbol{k}\in\mathbb{N}_0^d}\|_2.$$

Due to the ordering of the weights in (7.4), it is natural to approximate the solution using the first n series coefficients as follows:

$$\mathrm{APP}(f,n) := \sum_{i=1}^{n} \widehat{f}(\boldsymbol{k}_i) u_{\boldsymbol{k}_i} \quad \forall f \in \mathcal{F},\ n \in \mathbb{N}.$$

The error of this approximation in terms of the norm on the output space, $\mathcal{G}$, can be expressed as

$$\|\mathrm{SOL}(f) - \mathrm{APP}(f,n)\|_{\mathcal{G}} = \|(\widehat{f}(\boldsymbol{k}_i))_{i=n+1}^{\infty}\|_2.$$

Here, we have chosen $\tau = 2$, so $\rho' = \infty$. If one has a fixed data budget, n, then $\mathrm{APP}(f,n)$ is the best answer for certain classes of inputs in $\mathcal{F}$ as, for example, balls, which are among the most prominent examples studied in the literature.

We return to the example of functions on $[-1,1]^d$, using a Chebyshev polynomial basis, in Section 7.3.3 to discuss the information cost and computational complexity. We perform some numerical experiments for this example in Section 7.5.3.

7.2.3 Approximation with inputs lying in a ball

The optimal approximation based on n series coefficients of the input function is defined in terms of the series coefficients of the input function corresponding to the largest $\lambda_{\boldsymbol{k}}$ as follows:

$$\mathrm{APP}: \mathcal{F} \times \mathbb{N}_0 \to \mathcal{G}, \quad \mathrm{APP}(f,0) = 0, \quad \mathrm{APP}(f,n) := \sum_{i=1}^{n} \widehat{f}(\boldsymbol{k}_i) v_{\boldsymbol{k}_i} \ \forall n \in \mathbb{N}. \tag{7.11}$$

By the argument leading to (7.7), it follows that

$$\|\mathrm{SOL}(f) - \mathrm{APP}(f,n)\|_{\mathcal{G}} = \|(\widehat{f}(\boldsymbol{k}_i))_{i=n+1}^{\infty}\|_{\tau}. \tag{7.12}$$

An upper bound on the approximation error follows from Lemma 7.1:

$$\|\mathrm{SOL}(f) - \mathrm{APP}(f,n)\|_{\mathcal{G}} \le \left\| \left(\frac{\widehat{f}(\boldsymbol{k}_i)}{\lambda_{\boldsymbol{k}_i}} \right)_{i=n+1}^{\infty} \right\|_{\rho} \|\boldsymbol{\lambda}^{(n+1)}\|_{\rho'}, \tag{7.13}$$

where the tail of $\boldsymbol{\lambda}$ is defined by

$$\boldsymbol{\lambda}^{(n)} := (\lambda_{\boldsymbol{k}_i})_{i=n}^{\infty}. \tag{7.14}$$

This leads to the theorem below.

Theorem 7.2. *Let $\mathcal{B}_R := \{f \in \mathcal{F} : \|f\|_{\mathcal{F}} \le R\}$ denote the ball of radius R in the space of input functions. The error of the approximation defined in* (7.11) *is bounded tightly above as*

$$\sup_{f \in \mathcal{B}_R} \|\mathrm{SOL}(f) - \mathrm{APP}(f,n)\|_{\mathcal{G}} \le R \, \|\boldsymbol{\lambda}^{(n+1)}\|_{\rho'}. \tag{7.15}$$

Moreover, the worst-case error over $\mathcal{B}_R$ of $\mathrm{APP}'(\cdot,n)$*, for any approximation based on n series coefficients of the input function, can be no smaller.*

Proof. The proof of (7.15) follows immediately from (7.13) and Lemma 7.1. The optimality of APP follows by bounding the error of an arbitrary approximation, APP′, applied to functions that mimic the zero function.

Let $\mathrm{APP}'(0,n)$ depend on the series coefficients indexed by $\mathcal{J} = \{\boldsymbol{k}_1', \ldots, \boldsymbol{k}_n'\}$. Use Lemma 7.1 with $\mathcal{K} = \mathbb{K} \setminus \mathcal{J}$ to choose f to mimic the zero function, have norm R, and have as large a solution as possible by (7.6), i. e.,

$$\begin{gathered} \widehat{f}(\boldsymbol{k}_1') = \cdots = \widehat{f}(\boldsymbol{k}_n') = 0, \quad \|f\|_{\mathcal{F}} = R, \\ \|\mathrm{SOL}(f)\|_{\mathcal{G}} = R \|(\lambda_{\boldsymbol{k}})_{\boldsymbol{k} \notin \mathcal{J}}\|_{\rho'}. \end{gathered} \tag{7.16}$$

Then $\mathrm{APP}'(\pm f, n) = \mathrm{APP}'(0,n)$, because f mimics the zero function, and

$$\begin{aligned} &\sup_{f \in \mathcal{B}_R} \|\mathrm{SOL}(f) - \mathrm{APP}(f,n)\|_{\mathcal{G}} \\ &\quad \ge \max_{\pm} \|\mathrm{SOL}(\pm f) - \mathrm{APP}'(\pm f, n)\|_{\mathcal{G}} = \max_{\pm} \|\mathrm{SOL}(\pm f) - \mathrm{APP}'(0,n)\|_{\mathcal{G}} \end{aligned}$$

$$\geq \frac{1}{2}[\|\mathrm{SOL}(f) - \mathrm{APP}'(0,n)\|_{\mathcal{G}} + \|-\mathrm{SOL}(f) - \mathrm{APP}'(0,n)\|_{\mathcal{G}}]$$
$$\geq \|\mathrm{SOL}(f)\|_{\mathcal{G}} = R\|(\lambda_{\boldsymbol{k}})_{\boldsymbol{k}\notin\mathcal{J}}\|_{\rho'} \quad \text{by (7.16)}.$$

The ordering of the $\lambda_{\boldsymbol{k}}$ implies that $\|(\lambda_{\boldsymbol{k}})_{\boldsymbol{k}\notin\mathcal{J}}\|_{\rho'}$ for arbitrary $\mathcal{J}$ can be no smaller than the case $\mathcal{J} = \{\boldsymbol{k}_1, \dots, \boldsymbol{k}_n\}$. This completes the proof. □

Whereas approximation APP is a key piece of the puzzle, our ultimate goal is an algorithm, ALG : $\mathcal{C} \times [0, \infty)$, satisfying the absolute error criterion (7.1). The nonadaptive Algorithm 7.1 satisfies this error criterion for $\mathcal{C} = \mathcal{B}_R$.

Algorithm 7.1 Nonadaptive ALG for a ball of input functions.

Parameters: the Banach spaces $\mathcal{F}$ and $\mathcal{G}$, including the weights $\boldsymbol{\lambda}$; the ball radius, R; APP satisfying (7.15)
Input: a black-box function, f; an absolute error tolerance, $\varepsilon > 0$
Ensure: Error criterion (7.1) for $\mathcal{C} = \mathcal{B}_R$
Choose $n^* = \min\{n \in \mathbb{N}_0 : \|\boldsymbol{\lambda}^{(n+1)}\|_{\rho'} \leq \varepsilon/R\}$
Return $\mathrm{ALG}(f,\varepsilon) = \mathrm{APP}(f, n^*)$

We now show that this algorithm is optimal if the input set is a ball.

Theorem 7.3. *The nonadaptive Algorithm 7.1 has an information cost for the set of input functions $\mathcal{C} = \mathcal{B}_R$ that is given by*

$$\mathrm{COST}(\mathrm{ALG}, \mathcal{B}_R, \varepsilon, R') = \min\{n \in \mathbb{N}_0 : \|\boldsymbol{\lambda}^{(n+1)}\|_{\rho'} \leq \varepsilon/R\}, \quad R' \in (0, R].$$

This algorithm is essentially optimal for the set of input functions $\mathcal{B}_R$, namely,

$$\mathrm{COST}(\mathrm{ALG}, \mathcal{B}_R, \varepsilon, R') \leq \mathrm{COMP}(\mathcal{B}_R, \omega\varepsilon, R')$$
$$\forall \varepsilon \in (0, \varepsilon_{\max}],\ R' \in [R_{\min}, R],\ R \in [R_{\min}, \infty),$$

where $\varepsilon_{\max}$ and $R_{\min}$ are arbitrary and fixed, and $\omega = R_{\min}/R$.

Proof. Fix positive $\varepsilon_{\max}$, $R_{\min}$, R, and ω, as defined above. For $0 < \varepsilon \leq \varepsilon_{\max}$ and $R_{\min} \leq R' \leq R$, the information cost of nonadaptive Algorithm 7.1 follows from its definition. Let $n^*(\varepsilon, R) := \mathrm{COST}(\mathrm{ALG}, \mathcal{B}_R, \varepsilon, R')$. Construct an input function $f \in \mathcal{B}_{R'}$ as in the proof of Theorem 7.2 with $\mathcal{J} = \{\boldsymbol{k}_1, \dots, \boldsymbol{k}_{n^*(\omega\varepsilon, R')}\}$. By the argument in the proof of Theorem 7.2, any algorithm in $\mathcal{A}(\mathcal{B}_{R'})$ that can approximate SOL(f) with an error no greater than $\omega\varepsilon$ must use at least $n^*(\omega\varepsilon, R')$ series coefficients. Thus,

$$\begin{aligned}\mathrm{COST}(\mathrm{ALG}, \mathcal{B}_R, \varepsilon, R') &= n^*(\varepsilon, R) = n^*(\varepsilon R'/R, R') \\ &\leq n^*(\omega\varepsilon, R') \quad \text{since } R'/R \geq \omega \\ &\leq \mathrm{COMP}(\mathcal{B}_{R'}, \omega\varepsilon, R') \leq \mathrm{COMP}(\mathcal{B}_R, \omega\varepsilon, R').\end{aligned}$$

Thus, Algorithm 7.1 is essentially optimal. □

For Algorithm 7.1, the information cost, $\mathrm{COST}(\mathrm{ALG}, \mathcal{B}_R, \varepsilon, R)$, depends on the decay rate of the tail norm of the $\lambda_{\boldsymbol{k}_i}$. This decay may be algebraic or exponential, and also determines the problem complexity, $\mathrm{COMP}(\mathcal{B}_R, \varepsilon, R)$, as a function of the error tolerance, ε.

This theorem illustrates how an essentially optimal algorithm for solving a problem for a ball of input functions, $\mathcal{C} = \mathcal{B}_R$, can be nonadaptive. However, as alluded to above, we claim that it is impractical to know a priori which ball your input function lies in. On the other hand, in the situations described below, where $\mathcal{C}$ is a cone, we will show that $\mathcal{A}(\mathcal{C})$ actually contains only adaptive algorithms via the lemma below. The proof of this lemma follows directly from the definition of nonadaptivity.

Lemma 7.4. *For a given set of input functions, $\mathcal{C}$, if $\mathcal{A}(\mathcal{C})$ contains any nonadaptive algorithms, then for every $\varepsilon > 0$,*

$$\mathrm{COMP}(\mathcal{C}, \varepsilon) := \sup_{R>0} \mathrm{COMP}(\mathcal{C}, \varepsilon, R) < \infty.$$

7.3 Bounding the norm of the input function based on a pilot sample

7.3.1 The cone and the optimal algorithm

The premise of an adaptive algorithm is that the finite information we observe about the input function tells us something about what is not observed. Let n_1 denote the number of pilot observations, based on the set of wavenumbers

$$\mathcal{K}_1 := \{\boldsymbol{k}_1, \dots, \boldsymbol{k}_{n_1}\}, \tag{7.17}$$

where the $\boldsymbol{k}_i$ are defined by the ordering of the $\lambda_{\boldsymbol{k}}$ in (7.4). Let A be some constant inflation factor greater than one.

Accordingly, we now define the cone of functions, whose norm can be bounded well in terms of a pilot sample, $\{\widehat{f}(\boldsymbol{k}_1), \dots, \widehat{f}(\boldsymbol{k}_{n_1})\}$. It is given by

$$\mathcal{C}_{\boldsymbol{\lambda}, n_1, A} = \left\{ f \in \mathcal{F} : \|f\|_{\mathcal{F}} \leq A \left\| \left(\frac{\widehat{f}(\boldsymbol{k}_i)}{\lambda_{\boldsymbol{k}_i}} \right)_{i=1}^{n_1} \right\|_{\rho} \right\}. \tag{7.18}$$

Referring to error bound (7.13), we see that the error of $\mathrm{APP}(f, n)$ depends on the series coefficients not sampled. The definition of $\mathcal{C}_{\boldsymbol{\lambda}, n_1, A}$ allows us to bound these as follows:

$$\begin{aligned} &\left\| \left(\frac{\widehat{f}(\boldsymbol{k}_i)}{\lambda_{\boldsymbol{k}_i}} \right)_{i=n+1}^{\infty} \right\|_{\rho} \\ &= \left[\|f\|_{\mathcal{F}}^{\rho} - \left\| \left(\frac{\widehat{f}(\boldsymbol{k}_i)}{\lambda_{\boldsymbol{k}_i}} \right)_{i=1}^{n} \right\|_{\rho}^{\rho} \right]^{1/\rho} \qquad \forall f \in \mathcal{F} \end{aligned}$$

$$\le \left[A^\rho \left\| \left(\frac{\widehat{f}(\boldsymbol{k}_i)}{\lambda_{\boldsymbol{k}_i}} \right)_{i=1}^{n_1} \right\|_\rho^\rho - \left\| \left(\frac{\widehat{f}(\boldsymbol{k}_i)}{\lambda_{\boldsymbol{k}_i}} \right)_{i=1}^{n} \right\|_\rho^\rho \right]^{1/\rho} \quad \forall f \in \mathcal{C}_{\boldsymbol{\lambda},n_1,A}.$$

This inequality together with error bound (7.13) implies the data-based error bound

$$\|\mathrm{SOL}(f) - \mathrm{APP}(f,n)\|_{\mathcal{G}} \le \mathrm{ERR}\big((\widehat{f}(\boldsymbol{k}_i))_{i=1}^n, n\big) \quad \forall f \in \mathcal{C}_{\boldsymbol{\lambda},n_1,A}, \tag{7.19a}$$

where

$$\begin{aligned} &\mathrm{ERR}\big((\widehat{f}(\boldsymbol{k}_i))_{i=1}^n, n\big) \\ &\quad := \left[A^\rho \left\| \left(\frac{\widehat{f}(\boldsymbol{k}_i)}{\lambda_{\boldsymbol{k}_i}} \right)_{i=1}^{n_1} \right\|_\rho^\rho - \left\| \left(\frac{\widehat{f}(\boldsymbol{k}_i)}{\lambda_{\boldsymbol{k}_i}} \right)_{i=1}^{n} \right\|_\rho^\rho \right]^{1/\rho} \|\boldsymbol{\lambda}^{(n+1)}\|_{\rho'}, \\ &\forall n \ge n_1. \end{aligned} \tag{7.19b}$$

This error bound decays as $\|(f(\boldsymbol{k}_i)/\lambda_{\boldsymbol{k}_i})_{i=1}^n\|_\rho$ increases, and as the tail norm of the $\lambda_{\boldsymbol{k}_i}$ decreases. This data-driven error bound underlies Algorithm 7.2, which is successful for the cone of input functions, $\mathcal{C}_{\boldsymbol{\lambda},n_1,A}$, defined in (7.18):

Algorithm 7.2 ALG based on a pilot sample.

Parameters: the Banach spaces $\mathcal{F}$ and $\mathcal{G}$, including the weights $\boldsymbol{\lambda}$; an initial sample size, $n_1 \in \mathbb{N}$; an inflation factor, $A > 1$; APP satisfying (7.13)
Input: a black-box function, f; an absolute error tolerance, $\varepsilon > 0$
Ensure: Error criterion (7.1) for the cone defined in (7.18)
  Let $n \leftarrow n_1 - 1$
  repeat
    Let $n \leftarrow n + 1$
    Compute $\mathrm{ERR}\big((\widehat{f}(\boldsymbol{k}_i))_{i=1}^n, n\big)$ as defined in (7.19)
  until $\mathrm{ERR}\big((\widehat{f}(\boldsymbol{k}_i))_{i=1}^n, n\big) \le \varepsilon$
  Return $\mathrm{ALG}(f,\varepsilon) = \mathrm{APP}(f,n)$

Theorem 7.5. *Algorithm 7.2 yields an answer satisfying absolute error criterion* (7.1), *i. e.,* $\mathrm{ALG} \in \mathcal{A}(\mathcal{C}_{\boldsymbol{\lambda},n_1,A})$ *for* $\mathcal{C}_{\boldsymbol{\lambda},n_1,A}$, *defined in* (7.18). *The information cost is*

$$\begin{aligned} &\mathrm{COST}(\mathrm{ALG}, \mathcal{C}_{\boldsymbol{\lambda},n_1,A}, \varepsilon, R) \\ &\quad = \min\{n \ge n_1 : \|\boldsymbol{\lambda}^{(n+1)}\|_{\rho'} \le \varepsilon/[(A^\rho - 1)^{1/\rho} R]\}. \end{aligned} \tag{7.20}$$

There exist positive $\varepsilon_{\max}$ *and* $R_{\min}$, *for which the computational complexity has the lower bound*

$$\begin{aligned} &\mathrm{COMP}(\mathcal{C}_{\boldsymbol{\lambda},n_1,A}, \varepsilon, R) \ge \min\{n \ge n_1 : \|\boldsymbol{\lambda}^{(n+1)}\|_{\rho'} \le 2\varepsilon/[(1 - 1/A)R]\} \\ &\forall \varepsilon \in (0, \varepsilon_{\max}],\ R \in [R_{\min}, \infty). \end{aligned} \tag{7.21}$$

Algorithm 7.2 is essentially optimal. Moreover, $\mathcal{A}(\mathcal{C}_{\boldsymbol{\lambda},n_1,A})$ contains only adaptive algorithms.

Proof. The upper bound on the information cost of this algorithm is obtained by noting that

$$\begin{aligned}
&\operatorname{COST}(\text{ALG}, \mathcal{C}_{\boldsymbol{\lambda},n_1,A}, \varepsilon, R) \\
&= \max_{f \in \mathcal{C} \cap \mathcal{B}_R} \min\{n \geq n_1 : \operatorname{ERR}((\widehat{f}(\boldsymbol{k}_i))_{i=1}^n, n) \leq \varepsilon\} \\
&\leq \max_{f \in \mathcal{C} \cap \mathcal{B}_R} \min\left\{n \geq n_1 : (A^p - 1)^{1/p} \left\| \left(\frac{\widehat{f}(\boldsymbol{k}_i)}{\lambda_{\boldsymbol{k}_i}} \right)_{i=1}^{n_1} \right\|_p \|\boldsymbol{\lambda}^{(n+1)}\|_{p'} \leq \varepsilon \right\} \\
&\leq \min\{n \geq n_1 : (A^p - 1)^{1/p} R \,\|\boldsymbol{\lambda}^{(n+1)}\|_{p'} \leq \varepsilon\},
\end{aligned}$$

since $\|(\widehat{f}(\boldsymbol{k}_i)/\lambda_{\boldsymbol{k}_i})_{i=1}^{n_1}\|_p \leq \|(\widehat{f}(\boldsymbol{k}_i)/\lambda_{\boldsymbol{k}_i})_{i=1}^{n}\|_p \leq \|f\|_{\mathcal{F}} \leq R$ for all $f \in \mathcal{B}_R$, $n \geq n_1$. Moreover, this inequality is tight for some $f \in \mathcal{C}_{\boldsymbol{\lambda},n_1,A} \cap \mathcal{B}_R$, namely, those certain f for which $\widehat{f}(\boldsymbol{k}_i) = 0$ for $i > n_1$. This completes the proof of (7.20).

To prove the lower complexity bound, choose $\varepsilon_{\max}$ and $R_{\min}$ such that

$$\|(\lambda_{\boldsymbol{k}_i})_{i=n_1+1}^{\infty}\|_{p'} > 2\varepsilon_{\max}/[(1 - 1/A)R_{\min}].$$

Let ALG′ be any algorithm that satisfies the error criterion, (7.1), for this choice of $\mathcal{C}_{\boldsymbol{\lambda},n_1,A}$ in (7.18). Fix $R \in [R_{\min}, \infty)$ and $\varepsilon \in (0, \varepsilon_{\max}]$ arbitrarily. Two fooling functions will be constructed of the form $f_\pm = f_1 \pm f_2$.

The input function f_1 is defined via its series coefficients as in Lemma 7.1, having nonzero coefficients only for $\boldsymbol{k} \in \mathcal{K}_1 = \{\boldsymbol{k}_1, \ldots, \boldsymbol{k}_{n_1}\}$:

$$|\widehat{f}_1(\boldsymbol{k})| = \begin{cases} \dfrac{R(1+1/A)\lambda_{\boldsymbol{k}}^{p'/p+1}}{2\|(\lambda_{\boldsymbol{k}})_{\boldsymbol{k}\in\mathcal{K}_1}\|_{p'}^{p'/p}}, & \boldsymbol{k} \in \mathcal{K}_1, \\ 0, & \boldsymbol{k} \notin \mathcal{K}_1, \end{cases} \qquad \|f_1\|_{\mathcal{F}} = \frac{R(1+1/A)}{2}.$$

Suppose that $\text{ALG}'(f_1, \varepsilon)$ samples the series coefficients $\widehat{f}_1(\boldsymbol{k})$ for $\boldsymbol{k} \in \mathcal{J}$, and let n denote the cardinality of $\mathcal{J}$.

Now, construct the input function f_2, having zero coefficients for $\boldsymbol{k} \in \mathcal{J}$ and also as in Lemma 7.1:

$$\begin{aligned}
|\widehat{f}_2(\boldsymbol{k})| &= \begin{cases} \dfrac{R(1-1/A)\lambda_{\boldsymbol{k}}^{p'/p+1}}{2\|(\lambda_{\boldsymbol{k}})_{\boldsymbol{k}\notin\mathcal{J}}\|_{p'}^{p'/p}}, & \boldsymbol{k} \notin \mathcal{J}, \\ 0, & \boldsymbol{k} \in \mathcal{J}, \end{cases} \qquad \|f_2\|_{\mathcal{F}} = \frac{R(1-1/A)}{2}, \\
\|\operatorname{SOL}(f_2)\|_{\mathcal{G}} &= \frac{R(1-1/A)}{2} \|(\lambda_{\boldsymbol{k}})_{\boldsymbol{k}\notin\mathcal{J}}\|_{p'}.
\end{aligned} \tag{7.22}$$

Let $f_\pm = f_1 \pm f_2$. By the definitions above, it follows that

$$\|f_\pm\|_{\mathcal{F}} = \|f_1 \pm f_2\|_{\mathcal{F}} \leq \|f_1\|_{\mathcal{F}} + \|f_2\|_{\mathcal{F}} = R,$$

$$\begin{aligned}\left\|\left(\frac{\widehat{f}_{\pm}(\boldsymbol{k}_i)}{\lambda_{\boldsymbol{k}_i}}\right)_{i=1}^{n_1}\right\|_p &= \left\|\left(\frac{\widehat{f}_1(\boldsymbol{k}_i) \pm \widehat{f}_2(\boldsymbol{k}_i)}{\lambda_{\boldsymbol{k}_i}}\right)_{i=1}^{n_1}\right\|_p \\ &\geq \left\|\left(\frac{\widehat{f}_1(\boldsymbol{k}_i)}{\lambda_{\boldsymbol{k}_i}}\right)_{i=1}^{n_1}\right\|_p - \left\|\left(\frac{\widehat{f}_2(\boldsymbol{k}_i)}{\lambda_{\boldsymbol{k}_i}}\right)_{i=1}^{n_1}\right\|_p \\ &\geq \|f_1\|_{\mathcal{F}} - \|f_2\|_{\mathcal{F}} = \frac{R}{A} \geq \frac{\|f_{\pm}\|_{\mathcal{F}}}{A}.\end{aligned}$$

Therefore, $f_{\pm} \in \mathcal{C}_{\boldsymbol{\lambda},n_1,A} \cap \mathcal{B}_R$. Moreover, since the series coefficients for $f_{\pm}$ are the same for $\boldsymbol{k} \in \mathcal{J}$, it follows that $\mathrm{ALG}'(f_+,\varepsilon) = \mathrm{ALG}'(f_-,\varepsilon)$. Thus, $\mathrm{SOL}(f_+)$ must be quite similar to $\mathrm{SOL}(f_-)$.

Using an argument like that in the proof of Theorem 7.2, it follows that

$$\begin{aligned}\varepsilon &\geq \max_{\pm}\|\mathrm{SOL}(f_{\pm}) - \mathrm{ALG}'(f_{\pm},\varepsilon)\|_{\mathcal{G}} = \max_{\pm}\|\mathrm{SOL}(f_{\pm}) - \mathrm{ALG}'(f_{+},\varepsilon)\|_{\mathcal{G}} \\ &\geq \frac{1}{2}[\|\mathrm{SOL}(f_+) - \mathrm{ALG}'(f_+,\varepsilon)\|_{\mathcal{G}} + \|\mathrm{SOL}(f_-) - \mathrm{ALG}'(f_+,\varepsilon)\|_{\mathcal{G}}] \\ &\geq \frac{1}{2}\|\mathrm{SOL}(f_+ - f_-)\|_{\mathcal{G}} = \|\mathrm{SOL}(f_2)\|_{\mathcal{G}} = \frac{R(1-1/A)}{2}\|(\lambda_{\boldsymbol{k}})_{\boldsymbol{k}\notin\mathcal{J}}\|_{p'} \quad \text{by (7.22)} \\ &\geq \frac{R(1-1/A)}{2}\|\boldsymbol{\lambda}^{(n+1)}\|_{p'},\end{aligned}$$

by the ordering of the $\boldsymbol{k}$ in (7.4). By the choice of $R_{\min}$ and $\varepsilon_{\max}$ above, it follows that $n > n_1$. This inequality then implies lower complexity bound (7.21). Because $\lim_{R\to\infty} \mathrm{COMP}(\mathcal{C}_{\boldsymbol{\lambda},n_1,A},\varepsilon,R) = \infty$, it follows from Lemma 7.4 that $\mathcal{A}(\mathcal{C}_{\boldsymbol{\lambda},n_1,A})$ contains only adaptive algorithms.

The essential optimality of Algorithm 7.2 follows by observing that

$$\mathrm{COST}(\mathrm{ALG},\mathcal{C}_{\boldsymbol{\lambda},n_1,A},\varepsilon,R) \leq \mathrm{COMP}(\mathcal{C}_{\boldsymbol{\lambda},n_1,A},\omega\varepsilon,R) \quad \text{for } \omega = \frac{1-1/A}{2(A^p-1)^{1/p}}.$$

This satisfies definition (7.2). □

The above derivation assumes that $A > 1$. If $A = 1$, then our cone consists of functions, whose only nonzero series coefficients are for wavenumbers in the set $\{\boldsymbol{k}_1,\ldots,\boldsymbol{k}_{n_1}\}$. The exact solution can be constructed using only the pilot sample. Our algorithm is then nonadaptive, but succeeds for input functions in the cone $\mathcal{C}_{\boldsymbol{\lambda},n_1,A}$, which is an unbounded set.

We may not be able to guarantee that a particular f of interest lies in our cone, $\mathcal{C}_{\boldsymbol{\lambda},n_1,A}$, but the definition of $\mathrm{ERR}((\widehat{f}(\boldsymbol{k}_i))_{i=1}^n, n)$ in (7.19b) provides a necessary condition for f to lie in $\mathcal{C}_{\boldsymbol{\lambda},n_1,A}$, namely,

$$\left\|\left(\frac{\widehat{f}(\boldsymbol{k}_i)}{\lambda_{\boldsymbol{k}_i}}\right)_{i=1}^{n}\right\|_p \leq A\left\|\left(\frac{\widehat{f}(\boldsymbol{k})}{\lambda_{\boldsymbol{k}}}\right)_{\boldsymbol{k}\in\mathcal{K}_1}\right\|_p \quad \forall n \in \mathbb{N}. \tag{7.23}$$

If condition (7.23) is violated in practice, then $f \notin \mathcal{C}_{\boldsymbol{\lambda},n_1,A}$, and Algorithm 7.2 may output an incorrect answer. The remedy is to make $\mathcal{C}_{\boldsymbol{\lambda},n_1,A}$ more inclusive by increasing the inflation factor, A, and/or the pilot sample size, n_1.

Our $\mathcal{C}_{\boldsymbol{\lambda},n_1,A}$ is a cone, for which a pilot sample of the input function allows you to bound its norm. An alternative is to define a cone, for which a weaker norm of the function provides an upper bound on the original norm. Here is such a cone:

$$\begin{aligned} \mathcal{C}'_{\boldsymbol{\lambda},\boldsymbol{\omega},A'} &:= \{f \in \mathcal{F} : \|f\|_{\mathcal{F}} \le A'\|f\|_{\mathcal{F}'}\}, \\ \text{where } \|f\|_{\mathcal{F}'} &:= \left\| \left(\frac{\widehat{f}(\boldsymbol{k}_i)\omega_i}{\lambda_{\boldsymbol{k}_i}} \right)_{i=1}^{n} \right\|_p \text{ and } \omega_1 \ge \omega_2 \ge \cdots . \end{aligned} \tag{7.24}$$

The norm $\|\cdot\|_{\mathcal{F}'}$ is weaker than the norm $\|\cdot\|_{\mathcal{F}}$, so this cone is a strict subset of $\mathcal{F}$. The following proposition describes how the cones defined in (7.18) and (7.24) are related:

Proposition 7.6. *If the parameters A, A', A'', n_1, and $\boldsymbol{\omega} = (\omega_1, \omega_2, \ldots)$ satisfy*

$$\omega_1 A' \left[\frac{1 + (\omega_{n_1+1}/\omega_1)^p}{1 - (\omega_{n_1+1}A')^p} \right]^{1/p} \le A \le \omega_{n_1} A'' \textit{ and } \quad \omega_{n_1+1}A' < 1,$$

then the two types of cones defined in (7.18) and (7.24) include one another:

$$\mathcal{C}'_{\boldsymbol{\lambda},\boldsymbol{\omega},A'} \subseteq \mathcal{C}_{\boldsymbol{\lambda},n_1,A} \subseteq \mathcal{C}'_{\boldsymbol{\lambda},\boldsymbol{\omega},A''}.$$

Proof. Let us prove $\mathcal{C}_{\boldsymbol{\lambda},n_1,A} \subseteq \mathcal{C}'_{\boldsymbol{\lambda},\boldsymbol{\omega},A''}$ first. If $f \in \mathcal{C}_{\boldsymbol{\lambda},n_1,A}$, then by the definition of $\mathcal{C}_{\boldsymbol{\lambda},n_1,A}$, we have

$$\|f\|_{\mathcal{F}} \le A \left\| \left(\frac{\widehat{f}(\boldsymbol{k}_i)}{\lambda_{\boldsymbol{k}_i}} \right)_{i=1}^{n_1} \right\|_p \le A\omega_{n_1}^{-1} \left\| \left(\frac{\widehat{f}(\boldsymbol{k}_i)\omega_{\boldsymbol{k}_i}}{\lambda_{\boldsymbol{k}_i}} \right)_{i=1}^{n_1} \right\|_p \le A\omega_{n_1}^{-1} \|f\|_{\mathcal{F}'}.$$

Thus, if $A\omega_{n_1}^{-1} \le A''$, i. e. $A \le \omega_{n_1} A''$, then we must have $f \in \mathcal{C}'_{\boldsymbol{\lambda},\boldsymbol{\omega},A''}$. So, $\mathcal{C}_{\boldsymbol{\lambda},n_1,A} \subseteq \mathcal{C}'_{\boldsymbol{\lambda},\boldsymbol{\omega},A''}$.

For the proof that $\mathcal{C}'_{\boldsymbol{\lambda},\boldsymbol{\omega},A'} \subseteq \mathcal{C}_{\boldsymbol{\lambda},n_1,A}$, suppose $f \in \mathcal{C}'_{\boldsymbol{\lambda},\boldsymbol{\omega},A'}$. Then we have

$$\begin{aligned} \|f\|_{\mathcal{F}'}^p - \left\| \left(\frac{\widehat{f}(\boldsymbol{k}_i)\omega_i}{\lambda_{\boldsymbol{k}_i}} \right)_{i=1}^{n_1} \right\|_p^p &= \left\| \left(\frac{\widehat{f}(\boldsymbol{k}_i)\omega_i}{\lambda_{\boldsymbol{k}_i}} \right)_{i=n_1+1}^{\infty} \right\|_p^p \le \omega_{n_1+1}^p \left\| \left(\frac{\widehat{f}(\boldsymbol{k}_i)}{\lambda_{\boldsymbol{k}_i}} \right)_{i=n_1+1}^{\infty} \right\|_p^p \\ &= \omega_{n_1+1}^p \left[\|f\|_{\mathcal{F}}^p - \left\| \left(\frac{\widehat{f}(\boldsymbol{k}_i)}{\lambda_{\boldsymbol{k}_i}} \right)_{i=1}^{n_1} \right\|_p^p \right] \\ &\le \omega_{n_1+1}^p \left[A'^p \|f\|_{\mathcal{F}'}^p - \omega_1^{-p} \left\| \left(\frac{\widehat{f}(\boldsymbol{k}_i)\omega_i}{\lambda_{\boldsymbol{k}_i}} \right)_{i=1}^{n_1} \right\|_p^p \right]. \end{aligned}$$

Assuming $\omega_{n_1+1}A' < 1$ and rearranging the above terms, we obtain

$$\begin{aligned} \|f\|_{\mathcal{F}'}^p &\le \frac{1 + (\omega_{n_1+1}/\omega_1)^p}{1 - (\omega_{n_1+1}A')^p} \left\| \left(\frac{\widehat{f}(\boldsymbol{k}_i)\omega_i}{\lambda_{\boldsymbol{k}_i}} \right)_{i=1}^{n_1} \right\|_p^p, \\ \|f\|_{\mathcal{F}} &\le \omega_1 A' \left[\frac{1 + (\omega_{n_1+1}/\omega_1)^p}{1 - (\omega_{n_1+1}A')^p} \right]^{1/p} \left\| \left(\frac{\widehat{f}(\boldsymbol{k}_i)}{\lambda_{\boldsymbol{k}_i}} \right)_{i=1}^{n_1} \right\|_p. \end{aligned}$$

Hence, if $\omega_1 A' [\frac{1+(\omega_{n_1+1}/\omega_1)^p}{1-(\omega_{n_1+1}A')^p}]^{1/p} \le A$, it follows that $f \in \mathcal{C}_{\boldsymbol{\lambda},n_1,A}$, and so we obtain $\mathcal{C}'_{\boldsymbol{\lambda},\boldsymbol{\omega},A'} \subseteq \mathcal{C}_{\boldsymbol{\lambda},n_1,A}$. □

7.3.2 Bounds on the information cost for $\rho = \tau, \rho' = \infty$

We now derive upper and lower bounds on the information costs of Algorithms 7.1 and 7.2 for the case $\rho = \tau$ and $\rho' = \infty$, which includes the example introduced in Section 7.2.2. For $\rho' = \infty$, $\|\boldsymbol{\lambda}^{(n+1)}\|_{\rho'} = \lambda_{\boldsymbol{k}_{n+1}}$. To facilitate these bounds, we establish a couple of common lemmas. Again, we refer the reader to [10, Section 11.11].

Lemma 7.7. *Consider the ordered weights, $\lambda_{\boldsymbol{k}_1} \geq \lambda_{\boldsymbol{k}_2} \geq \cdots > 0$, and recall from (7.14) that $\boldsymbol{\lambda}^{(n)} = (\lambda_{\boldsymbol{k}_n}, \lambda_{\boldsymbol{k}_{n+1}}, \dots)$. It follows that*

$$\lambda_{\boldsymbol{k}_{n+1}} \leq \frac{\|\boldsymbol{\lambda}\|_p}{n^{1/p}} \quad \text{and} \quad \sup_{n \geq n_{\min}} \lambda_{\boldsymbol{k}_n} n^{1/p} \geq \frac{\|\boldsymbol{\lambda}^{(n_{\min})}\|_{p\eta}}{\zeta(\eta)^{1/(p\eta)}} \quad \forall p > 0,\ \eta > 1,\ n_{\min} \in \mathbb{N}.$$

If $\|\boldsymbol{\lambda}\|_p = \infty$ or $\sup_{n \geq n_{\min}} \lambda_{\boldsymbol{k}_n} n^{1/p} = \infty$, we consider these inequalities to be trivially true.

Proof. Since the $\lambda_{\boldsymbol{k}_i}$ are nonincreasing, it follows that

$$n\lambda_{\boldsymbol{k}_{n+1}}^p \leq \lambda_{\boldsymbol{k}_1}^p + \lambda_{\boldsymbol{k}_2}^p + \cdots + \lambda_{\boldsymbol{k}_n}^p \leq \sum_{\boldsymbol{k} \in \mathbb{K}} \lambda_{\boldsymbol{k}}^p = \|\boldsymbol{\lambda}\|_p^p \quad \forall p > 0.$$

Dividing both sides by n and taking the pth root completes the proof of the first inequality.

To prove the second inequality, we employ Hölder's inequality to show that

$$\|\boldsymbol{\lambda}^{(n_{\min})}\|_{p\eta}^{p\eta} = \sum_{n=n_{\min}}^{\infty} \frac{\lambda_{\boldsymbol{k}_n}^{p\eta} n^{\eta}}{n^{\eta}} \leq \sup_{n \geq n_{\min}} \lambda_{\boldsymbol{k}_n}^{p\eta} n^{\eta} \sum_{n=n_{\min}}^{\infty} \frac{1}{n^{\eta}} \leq \sup_{n \geq n_{\min}} \lambda_{\boldsymbol{k}_n}^{p\eta} n^{\eta} \zeta(\eta),$$

where ζ is the Riemann zeta function. The proof is completed by taking the $p\eta^{\text{th}}$ root and rearranging the terms. □

These upper and lower bounds together with Theorems 7.3 and 7.5 can be used to relate the information costs of Algorithms 7.1 and 7.2 to the decay rate of the elements of $\boldsymbol{\lambda}$.

Theorem 7.8. *If the elements of $\boldsymbol{\lambda}$ decay fast enough so that $\|\boldsymbol{\lambda}\|_p < \infty$ and $\rho' = \infty$, then the information costs of Algorithms 7.1 and 7.2 are bounded above by the following:*

$$\text{Algorithm 7.1:} \quad \operatorname{COST}(\text{ALG}, \mathcal{B}_R, \varepsilon, R') \leq \left\lceil \left(\frac{R\|\boldsymbol{\lambda}\|_p}{\varepsilon} \right)^p \right\rceil, \quad \forall \varepsilon,\ 0 < R' < R,$$

$$\text{Algorithm 7.2:} \quad \operatorname{COST}(\text{ALG}, \mathcal{C}_{\boldsymbol{\lambda}, n_1, A}, \varepsilon, R) \leq \max\left(n_1, \left\lceil \left(\frac{[(A^\rho - 1)^{1/\rho} R] \|\boldsymbol{\lambda}\|_p}{\varepsilon} \right)^p \right\rceil \right).$$

Conversely, for fixed R', R with $0 < R' < R$, if either

$$\operatorname{COST}(\text{ALG}, \mathcal{B}_R, \varepsilon, R') \leq C\varepsilon^{-p} \quad \forall \varepsilon \in (0, \varepsilon_{\max}] \quad \text{for Algorithm 7.1, or}$$

$$\mathrm{COST}(\mathrm{ALG}, \mathcal{C}_{\boldsymbol{\lambda},n_1,A}, \varepsilon, R) \le C\varepsilon^{-p} \quad \forall \varepsilon \in (0, \varepsilon_{\max}] \quad \textit{for Algorithm 7.2},$$

then $\|\boldsymbol{\lambda}\|_{p\eta} < \infty$ *for all* $\eta > 1$. *Since Algorithms 7.1 and 7.2 are essentially optimal, the above inequalities provide upper and lower asymptotic bounds on* $\mathrm{COMP}(\mathcal{B}_R, \varepsilon, R')$ *and* $\mathrm{COMP}(\mathcal{C}_{\boldsymbol{\lambda},n_1,A}, \varepsilon, R)$.

Proof. We give the proof for Algorithm 7.2, which is similar to the proof for Algorithm 7.1. Recall that for $\rho' = \infty$, $\|\boldsymbol{\lambda}^{(n+1)}\|_{\rho'} = \lambda_{\boldsymbol{k}_{n+1}}$. From (7.20) it follows that the information cost for Algorithm 7.2 is

$$\begin{aligned}
\mathrm{COST}(\mathrm{ALG}, \mathcal{C}_{\boldsymbol{\lambda},n_1,A}, \varepsilon, R) &= \min\left\{n \ge n_1 : \lambda_{\boldsymbol{k}_{n+1}} \le \frac{\varepsilon}{[(A^\rho - 1)^{1/\rho} R]}\right\} \\
&\le \min\left\{n \ge n_1 : \|\boldsymbol{\lambda}\|_p n^{-1/p} \le \frac{\varepsilon}{[(A^\rho - 1)^{1/\rho} R]}\right\} \\
&\qquad \text{by Lemma 7.7 since } \lambda_{\boldsymbol{k}_{n+1}} \le \|\boldsymbol{\lambda}\|_p n^{-1/p} \\
&= \left\lceil \left(\frac{[(A^\rho - 1)^{1/\rho} R] \|\boldsymbol{\lambda}\|_p}{\varepsilon} \right)^p \right\rceil .
\end{aligned}$$

Now, suppose that $\mathrm{COST}(\mathrm{ALG}, \mathcal{C}_{\boldsymbol{\lambda},n_1,A}, \varepsilon, R) \le C\varepsilon^{-p}$ for all $\varepsilon \in (0, \varepsilon_{\max}]$. Since the information cost is an integer, this means that $\mathrm{COST}(\mathrm{ALG}, \mathcal{C}_{\boldsymbol{\lambda},n_1,A}, \varepsilon, R) \le \lfloor C\varepsilon^{-p} \rfloor$ and let $n_{\min} = \lfloor C\varepsilon_{\max}^{-p} \rfloor + 1$. According to Lemma 7.7, it follows that for all $\eta > 1$,

$$\begin{aligned}
\|\boldsymbol{\lambda}^{(n_{\min})}\|_{p\eta} &\le \zeta(\eta)^{1/(p\eta)} \sup_{n \ge n_{\min}} \lambda_{\boldsymbol{k}_n} n^{1/p} \\
&= \zeta(\eta)^{1/(p\eta)} \sup_{0 < \varepsilon \le \varepsilon_{\max}} \lambda_{\boldsymbol{k}_{\lfloor C\varepsilon^{-p} \rfloor + 1}} (\lfloor C\varepsilon^{-p} \rfloor + 1)^{1/p} \\
&\qquad \text{by the definition of } n_{\min} \\
&\le \zeta(\eta)^{1/(p\eta)} \sup_{0 < \varepsilon \le \varepsilon_{\max}} \frac{\varepsilon\, (\lfloor C\varepsilon^{-p} \rfloor + 1)^{1/p}}{AR} < \infty \\
&\qquad \text{since } \mathrm{COST}(\mathrm{ALG}, \mathcal{C}_{\boldsymbol{\lambda},n_1,A}, \varepsilon, R) \le \lfloor C\varepsilon^{-p} \rfloor .
\end{aligned}$$

Since $\|\boldsymbol{\lambda}^{(n_{\min})}\|_{p\eta}$ is finite, it follows that $\|\boldsymbol{\lambda}\|_{p\eta}$ is finite. □

To be more concrete, we consider the situation where the $\lambda_{\boldsymbol{k}}$ are specified in terms of positive *coordinate weights*, $\boldsymbol{w} = (w_1, \ldots, w_d)$, and positive *smoothness weights*, $\boldsymbol{s} = (s_1, s_2, \ldots)$:

$$\lambda_{\boldsymbol{k}} := \prod_{\substack{\ell=1 \\ k_\ell > 0}}^{d} w_\ell s_{k_\ell}, \quad \boldsymbol{k} \in \mathbb{K} := \mathbb{N}_0^d,\ d \in \mathbb{N}. \tag{7.25}$$

This is a generalization of our illustrative example, (7.10), in Section 7.2.2, where $\boldsymbol{s} = (1, 2^{-r}, 3^{-r}, \ldots)$. This form of the $\lambda_{\boldsymbol{k}}$ is also considered in Sections 7.4 and 7.5.

Lemma 7.9. *For $\boldsymbol{\lambda} = (\lambda_{\boldsymbol{k}})_{\boldsymbol{k}\in\mathbb{N}_0^d}$, defined in (7.25), we have the following lower and upper bounds on $\|\boldsymbol{\lambda}\|_p$. It holds that*

$$\|\boldsymbol{s}\|_p\|\boldsymbol{w}\|_p \le \|\boldsymbol{\lambda}\|_p \le \exp\left(\frac{1}{p}\|\boldsymbol{s}\|_p^p\|\boldsymbol{w}\|_p^p\right) \quad \text{for general } \boldsymbol{s},$$

$$[\zeta(pr)]^{1/p}\|\boldsymbol{w}\|_p \le \|\boldsymbol{\lambda}\|_p \le \exp\left(\frac{\zeta(pr)}{p}\|\boldsymbol{w}\|_p^p\right) \quad \text{for } \boldsymbol{s} = (1, 2^{-r}, \dots),\ pr > 1.$$

Proof. The proof is constructed first for general $\boldsymbol{s}$. For the particular form of $\boldsymbol{s}$ introduced in (7.10), it follows that $\|\boldsymbol{s}\|_p^p = \zeta(pr)$ for $pr > 1$.

By straightforward calculation and (7.25),

$$\|\boldsymbol{\lambda}\|_p^p = \sum_{\boldsymbol{k}\in\mathbb{N}_0^d} \lambda_{\boldsymbol{k}}^p = \prod_{\ell=1}^{d}\left(1 + w_\ell^p \sum_{k=1}^{\infty} s_k^p\right) = \prod_{\ell=1}^{d}(1 + w_\ell^p\|\boldsymbol{s}\|_p^p). \tag{7.26}$$

A lower bound is obtained by noting from the equation above that

$$\|\boldsymbol{\lambda}\|_p^p = \prod_{\ell=1}^{d}(1 + w_\ell^p\|\boldsymbol{s}\|_p^p) \ge \|\boldsymbol{s}\|_p^p \sum_{\ell=1}^{d} w_\ell^p = \|\boldsymbol{s}\|_p^p\|\boldsymbol{w}\|_p^p.$$

Taking the pth root completes the proof of the lower bound.

For the upper bound (7.26) implies that

$$\begin{aligned}\|\boldsymbol{\lambda}\|_p^p &= \prod_{\ell=1}^{d}(1 + w_\ell^p\|\boldsymbol{s}\|_p^p) = \exp\left(\sum_{\ell=1}^{d}\log(1 + w_\ell^p\|\boldsymbol{s}\|_p^p)\right)\\ &\le \exp\left(\sum_{\ell=1}^{d} w_\ell^p\|\boldsymbol{s}\|_p^p\right) = \exp(\|\boldsymbol{w}\|_p^p\|\boldsymbol{s}\|_p^p) \quad \text{since } \log(1+x) \le x\ \forall x \ge 0.\end{aligned}$$

Again, taking the pth root completes the proof for general $\boldsymbol{s}$. □

Theorem 7.8 and Lemma 7.9 imply bounds on the information costs of Algorithms 7.1 and 7.2 for the function spaces defined in the illustrative example in Section 7.2.2.

Corollary 7.10. *For the function spaces defined in the illustrative example in Section 7.2.2, for all $p > 1/r$:*

$$\textit{Algorithm 7.1:} \quad \mathrm{COST}(\mathrm{ALG}, \mathcal{B}_R, \varepsilon, R') \le \left\lceil \left(\frac{R}{\varepsilon}\right)^p \exp(\zeta(pr)\|\boldsymbol{w}\|_p^p) \right\rceil,$$

$$\begin{aligned}\textit{Algorithm 7.2:} \quad &\mathrm{COST}(\mathrm{ALG}, \mathcal{C}_{\boldsymbol{\lambda},n_1,A}, \varepsilon, R)\\ &\le \max\left(n_1, \left\lceil \left(\frac{(A^p-1)^{1/p}R}{\varepsilon}\right)^p \exp(\zeta(pr)\|\boldsymbol{w}\|_p^p) \right\rceil\right).\end{aligned}$$

Furthermore, these information costs cannot grow any slower than $\mathcal{O}(\varepsilon^{-1/r})$ for $\varepsilon \downarrow 0$. Since Algorithms 7.1 and 7.2 are essentially optimal, the above inequalities provide upper and lower asymptotic bounds on $\mathrm{COMP}(\mathcal{B}_R, \varepsilon, R')$ and $\mathrm{COMP}(\mathcal{C}_{\boldsymbol{\lambda},n_1,A}, \varepsilon, R)$ for the illustrative example in Section 7.2.2.

7.3.3 Tractability

Besides understanding the dependence of COMP($\mathcal{C}, \varepsilon, R$) on ε, we also want to understand how COMP($\mathcal{C}, \varepsilon, R$) depends on the dimension of the domain of the input function. Suppose that $f : \Omega^d \to \mathbb{R}$, for some $\Omega \subseteq \mathbb{R}$, and let $\mathcal{F}_d$ denote the dependence of the input space on the dimension d. The set of functions, for which our algorithms succeed, depends on the dimension, too. Thus, for the sake of brevity, we write $\mathcal{C}_d$ instead of $\mathcal{C}_{\lambda,n_1,A}$ in this section. Also, SOL, APP, COST, and COMP depend implicitly on the dimension, and this dependence is sometimes indicated explicitly by the subscript d.

Different dependencies of COMP($\mathcal{C}_d, \varepsilon, R$) on the dimension d and the error tolerance ε are formalized as different notions of tractability. Since the complexity is defined in terms of the best algorithm, tractability is a property that is inherent to the problem, not to a particular algorithm. We define the following notions of tractability (for further information on tractability we refer to the trilogy [7–9]). Note that in contrast to these references, we explicitly include the dependence on R in our definitions. This dependence is natural for cones $\mathcal{C}$ and might be different if $\mathcal{C}$ is not a cone.

- We say that the adaptive approximation problem is strongly polynomially tractable if and only if there are nonnegative C, $\tilde{p}$, and positive $\varepsilon_{\max}$, and $R_{\min}$ such that

$$\mathrm{COMP}(\mathcal{C}_d, \varepsilon, R) \leq C\, R^{\tilde{p}}\, \varepsilon^{-\tilde{p}} \quad \forall d \in \mathbb{N},\ \varepsilon \in (0, \varepsilon_{\max}],\ R \in [R_{\min}, \infty).$$

  The infimum of $\tilde{p}$ satisfying the bound above is denoted by $\tilde{p}_*$ and is called the exponent of strong polynomial tractability.
- We say that the problem is polynomially tractable if and only if there are nonnegative C, $\tilde{p}$, q, and positive $\varepsilon_{\max}$, and $R_{\min}$ such that

$$\mathrm{COMP}(\mathcal{C}_d, \varepsilon, R) \leq C\, d^q\, R^{\tilde{p}}\, \varepsilon^{-\tilde{p}} \quad \forall d \in \mathbb{N},\ \varepsilon \in (0, \varepsilon_{\max}],\ R \in [R_{\min}, \infty).$$

- We say that the problem is weakly tractable iff

$$\lim_{d+R\varepsilon^{-1}\to\infty} \frac{\log \mathrm{COMP}(\mathcal{C}_d, \varepsilon, R)}{d + R\varepsilon^{-1}} = 0.$$

Necessary and sufficient conditions on these tractability notions will be studied for different types of algorithms in this section and Section 7.4.4.

We remark that, for the sake of brevity, we focus here on tractability notions that are summarized as algebraic tractability in the recent literature (see, e. g., [5]). Theoretically, one could also study exponential tractability, where one would essentially replace ε^{-1} by $\log(1 + \varepsilon^{-1})$ in the previous tractability notions. A more detailed study of tractability will be done in a future paper.

Returning again to the example in Section 7.2.2, we have, by Corollary 7.10

$$\mathrm{COMP}(\mathcal{B}_R, \varepsilon, R) \le \left\lceil \left(\frac{R}{\varepsilon}\right)^p \exp(\zeta(pr)\|\boldsymbol{w}\|_p^p) \right\rceil.$$

If $p^\dagger$ is the infimum of the p, for which $\|\boldsymbol{w}\|_p$ is finite, and $p^\dagger$ is finite, then we obtain strong polynomial tractability and an exponent of strong tractability that is $\widetilde{p}_* = \max(1/r, p^\dagger)$. On the other hand, if the coordinate weights are all unity, $w_1 = w_2 = \cdots = 1$, then there are 2^d different $\lambda_{\boldsymbol{k}}$ with a value of 1, and so $\mathrm{COMP}(\mathcal{B}_R, \varepsilon, R) \ge 2^d$, and the problem is not tractable.

Recall that we assume that $\lambda_{d,\boldsymbol{k}_1} \ge \lambda_{d,\boldsymbol{k}_2} \ge \cdots > 0$. Let

$$n_{\mathrm{COMP}}(\delta, d) := \min\{n \ge 0 : \|\boldsymbol{\lambda}_d^{(n+1)}\|_{\rho'} \le \delta\} \quad \forall \delta > 0,$$

where, as in (7.14), $\boldsymbol{\lambda}_d^{(n)} := (\lambda_{d,\boldsymbol{k}_n}, \lambda_{d,\boldsymbol{k}_{n+1}}, \ldots)$ From equations (7.20) and (7.21), we obtain that

$$\begin{aligned} &\mathrm{COMP}(\mathcal{C}_d, \omega_{\mathrm{lo}}\varepsilon, R) \le n_{\mathrm{COMP}}(\varepsilon/R, d) \le \mathrm{COMP}(\mathcal{C}_d, \omega_{\mathrm{hi}}\varepsilon, R) \\ &\forall \varepsilon \in (0, \varepsilon_{\max}],\ R \in [R_{\min}, \infty), \end{aligned}$$

where the positive constants ω_{lo} and ω_{hi} depend on A, but do not depend on d, ε, or R. From the equation above, it is clear that tractability depends on the behavior of $n_{\mathrm{COMP}}(\varepsilon/R, d)$ as R/ε and d tend to infinity. We would like to study under which conditions we obtain the various tractability notions defined above.

To this end, we distinguish two cases, depending on whether ρ' is infinite or not. This distinction is useful, because it allows us to relate the computational complexity of the algorithms considered in this chapter to the computational complexity of linear problems on certain function spaces considered in the classical literature on information-based complexity, as for example [7]. The case $\rho' = \infty$ corresponds to the worst-case setting, where one studies the worst performance of an algorithm over the unit ball of a space. The results in Theorem 7.11 below are indeed very similar to the results for the worst-case setting over balls of suitable function spaces. The case $\rho < \infty$ corresponds to the so-called average-case setting, where one considers the average performance over a function space equipped with a suitable measure. For both of these settings, there exist tractability results that we will make use of here.

7.3.3.1 CASE 1: $\rho' = \infty$

If $\rho' = \infty$, we have, due to the monotonicity of the $\lambda_{d,\boldsymbol{k}_i}$,

$$n_{\mathrm{COMP}}(\varepsilon/R, d) = \min\{n \ge 0 \colon \lambda_{d,\boldsymbol{k}_{n+1}} \le \varepsilon/R\}.$$

We then have the following theorem.

Theorem 7.11. *Using the same notation as above, the following statements hold for the case $\rho' = \infty$.*

1. *We have strong polynomial tractability if and only if there exist $p > 0$ and $i_0 \in \mathbb{N}$ such that*

$$\sup_{d\in\mathbb{N}} \|\boldsymbol{\lambda}_d^{(i_0)}\|_p < \infty. \tag{7.27}$$

   Furthermore, the exponent of strong polynomial tractability is then equal to the infimum of those $p > 0$ for which (7.27) holds.

2. *We have polynomial tractability if and only if there exist $p_1, p_2 \geq 0$ and $p_3, K > 0$ such that*

$$\sup_{d\in\mathbb{N}} d^{-p_1} \|\boldsymbol{\lambda}_d^{(\lceil Kd^{p_2}\rceil)}\|_{p_3} < \infty.$$

3. *We have weak tractability if and only if*

$$\sup_{d\in\mathbb{N}} \exp(-cd) \sum_{i=1}^{\infty} \exp\left(-c\left(\frac{1}{\lambda_{d,\mathbf{k}_i}}\right)\right) < \infty \quad \textit{for all} \quad c > 0. \tag{7.28}$$

Proof. If $\tilde{\varepsilon} := \sqrt{\varepsilon/R}$, then $n_{\mathrm{COMP}}(\varepsilon/R, d) = \min\{n \geq 0 : \lambda_{d,\mathbf{k}_{n+1}} \leq \tilde{\varepsilon}^2\}$. The latter expression is well studied in the context of tractability of linear problems in the worst-case setting defined on unit balls of certain spaces, and if and only if conditions on the $\lambda_{d,\mathbf{k}_i}$ for various tractability notions are known. These conditions can be found in [7, Chapter 5] for (strong) polynomial tractability, and [14] for weak tractability.

Since, in this chapter, we consider $\min\{n \geq 0 : \lambda_{d,\mathbf{k}_{n+1}} \leq \varepsilon/R\}$, and in [7] and [14] ε/R is replaced by the square of the error tolerance, there are slight differences between the results here and those in the aforementioned references; to be more precise, the exponent of strong polynomial tractability here is, formally speaking, half of the exponent in [7], and $1/\lambda_{d,\mathbf{k}_i}$ in (7.28) corresponds to $1/\sqrt{\lambda_{d,\mathbf{k}_i}}$ in [14]. □

7.3.3.2 CASE 2: $\rho' < \infty$

In this case, letting $\tilde{\varepsilon} := (\varepsilon/R)^{\rho'/2}$ and $\tilde{\lambda}_{d,i} := \lambda_{d,\mathbf{k}_i}^{\rho'}$, we have

$$\begin{aligned} n_{\mathrm{COMP}}(\varepsilon/R, d) &= \min\{n \geq 0 : \|\boldsymbol{\lambda}_d^{(n+1)}\|_{\rho'}^{\rho'} \leq (\varepsilon/R)^{\rho'}\} \\ &= \min\{n \geq 0 : \|\tilde{\boldsymbol{\lambda}}_d^{(n+1)}\|_1 \leq \tilde{\varepsilon}^2\}. \end{aligned} \tag{7.29}$$

However, the latter expression corresponds exactly to the average-case tractability (with respect to the parameters $\tilde{\lambda}_{d,i}$ and $\tilde{\varepsilon}$) defined on certain spaces as studied in, e. g., [7]. This leads us to the following theorem:

Theorem 7.12. *Using the same notation as above, the following statements hold for the case $\rho' < \infty$:*

1. *We have strong polynomial tractability if and only if there exist $p \in (0,1)$ and $i_0 \in \mathbb{N}$ such that*

$$\sup_{d\in\mathbb{N}} \|\boldsymbol{\lambda}_d^{(i_0)}\|_{\rho' p} < \infty. \tag{7.30}$$

Furthermore, the exponent of strong polynomial tractability is then

$$\inf\{\rho' p/(1-p) : p \text{ satisfies } (7.30)\}.$$

2. *We have polynomial tractability if and only if there exist $p_1, p_2 \geq 0$ and $p_3 \in (0,1)$, $K > 0$ such that*

$$\sup_{d\in\mathbb{N}} d^{-p_1} \|\boldsymbol{\lambda}_d^{(\lceil K d^{p_2}\rceil)}\|_{\rho' p_3} < \infty.$$

3. *We have weak tractability if and only if*

$$\lim_{n\to\infty} (\log n)^2 \|\boldsymbol{\lambda}_d^{(n)}\|_1 = 0 \quad \text{for all } d,$$

and there exists a function $f : [0,1/2) \to \{1,2,3,\ldots\}$ such that

$$\sup_{\beta\in(0,1/2]} \beta^{-2} \sup_{d\geq f(\beta)} \sup_{n\geq\lceil \exp(d\sqrt{b})\rceil+1} (\log n)^2 \|\boldsymbol{\lambda}_d^{(n)}\|_1 < \infty.$$

Proof. The proof of the theorem is similar to that of Theorem 7.11, using (7.29). □

Remark 7.1. Results for further tractability notions, such as quasipolynomial tractability or (s,t)-weak tractability, can be shown using similar arguments as above and results from [5, 8, 14], and the papers cited therein.

To be more concrete, we return to the situation where the $\lambda_{d,\boldsymbol{k}}$ are given as in (7.25) one more time. We note that $\|\boldsymbol{\lambda}_d^{(i_0)}\|_p^p < \infty$ holds if and only if $\|\boldsymbol{\lambda}_d\|_p^p < \infty$. We denote the d-dependence of the vector of coordinate weights by writing $\boldsymbol{w}_d := (w_1,\ldots,w_d)$ and $\boldsymbol{w}_\infty = (w_1, w_2, \ldots)$. By Lemma 7.9, we have

$$\|\boldsymbol{s}\|_p \|\boldsymbol{w}_d\|_p \leq \|\boldsymbol{\lambda}_d\|_p \leq \exp\left(\frac{1}{p}\|\boldsymbol{s}\|_p^p \|\boldsymbol{w}_d\|_p^p\right),$$

and this estimate holds for any $d \in \mathbb{N}$. Thus, we have necessary and sufficient conditions for strong tractability.

Corollary 7.13. *For the $\lambda_{d,\boldsymbol{k}}$ of the form (7.25), we have strong polynomial tractability if and only if there exists $p > 0$ such that*

$$\|\boldsymbol{s}\|_p < \infty \quad \text{and} \quad \|\boldsymbol{w}_\infty\|_p < \infty.$$

In particular, considering example in Section 7.2.2, where $s_k = k^{-r}$, we have strong polynomial tractability if and only if there exists $p > 0$ such that

$$rp > 1 \quad \textit{and} \quad \|\boldsymbol{w}_\infty\|_p < \infty.$$

Remark 7.2. Note that in the setting of this example, the term $\|\boldsymbol{\lambda}_d\|_p^p$ will usually depend exponentially on d unless the coordinate weights, w_ℓ, decay to zero fast enough with increasing ℓ. Hence, we can only hope for tractability under the presence of decaying w_ℓ. For further details on weighted approximation problems and tractability, we refer to [7].

7.4 Tracking the decay rate of the series coefficients of the input function

From error bound (7.12), it follows that the faster the $\hat{f}(\boldsymbol{k}_i)$ decay, the faster APP(f, n) converges to the solution. Unfortunately, adaptive Algorithm 7.2 does not adapt to the decay rate of the $\hat{f}(\boldsymbol{k}_i)$ as $i \to \infty$. It simply bounds $\|f\|_{\mathcal{F}}$ based on a pilot sample. The algorithm presented in this section tracks the rate of decay of the $\hat{f}(\boldsymbol{k}_i)$ and terminates sooner if the $\hat{f}(\boldsymbol{k}_i)$ decay more quickly. Similar algorithms for quasi-Monte Carlo integration are developed in [2, 4], and [3].

There is an implicit assumption in this section that function data are cheap, and we can afford a large sample size. A large sample size is required to do meaningful tracking of the decay of the series coefficients. The previous section and the next section are more suited to the case when function data are expensive and the final sample size must be modest.

Let $\boldsymbol{n} = (n_j)_{j\geq 0}$ be a strictly increasing sequence of nonnegative integers. This sequence may increase geometrically or algebraically. Define the sets of wavenumbers analogously to (7.17),

$$n_{-1} = 0, \quad \mathcal{K}_j := \{\boldsymbol{k}_{n_{j-1}+1}, \dots, \boldsymbol{k}_{n_j}\} \quad \text{for } j \in \mathbb{N}_0.$$

If $n_0 = 0$, then $\mathcal{K}_0$ is empty. For any $f \in \mathcal{F}$, define the norms of subsets of series coefficients:

$$\sigma_j(f) := \left\| \left(\frac{\hat{f}(\boldsymbol{k})}{\lambda_{\boldsymbol{k}}} \right)_{\boldsymbol{k}\in\mathcal{K}_j} \right\|_p \quad \text{for } j \in \mathbb{N}. \tag{7.31}$$

Thus, $\|f\|_{\mathcal{F}} = \|(\sigma_j(f))_{j\in\mathbb{N}_0}\|_p$.

As stated above, we would like to keep track of the rate of decay of the $\hat{f}(\boldsymbol{k}_i)$. Hence, for this section, we define the cone of input functions by

$$\mathcal{C}_{\boldsymbol{\lambda},\boldsymbol{n},a,b} := \{f \in \mathcal{F} : \sigma_{j+l}(f) \leq ab^l \sigma_j(f)\ \forall j, l \in \mathbb{N}\}. \tag{7.32}$$

Here, a and b are positive reals with $b < 1 < a$. The constant a is an inflation factor, and the constant b defines the general rate of decay of the $\sigma_j(f)$ for $f \in \mathcal{C}_{\boldsymbol{\lambda},\boldsymbol{n},a,b}$. We remark that knowing $\sigma_1(f)$ would already be sufficient to identify a sample size that guarantees an error of at most ε. However, knowing $\sigma_j(f)$ for further values of j will allow us to stop earlier.

Because ab^l may be greater than one, we do not require the series coefficients of the solution, SOL(f), to decay monotonically. However, we expect their partial sums to decay *steadily*. The series coefficients for wavenumbers $\boldsymbol{k} \in \mathcal{K}_0$ do not affect the definition of $\mathcal{C}_{\boldsymbol{\lambda},\boldsymbol{n},a,b}$, and may behave erratically. Lemma 7.1 implies that

$$\|(\widehat{f}(\boldsymbol{k}))_{\boldsymbol{k}\in\mathcal{K}_j}\|_\tau \le \sigma_j(f)\Lambda_j, \quad \text{where } \Lambda_j := \|(\lambda_{\boldsymbol{k}})_{\boldsymbol{k}\in\mathcal{K}_j}\|_{\rho'}. \tag{7.33}$$

From (7.8) and (7.9), it follows that the norm of the solution operator is

$$\|\mathrm{SOL}\|_{\mathcal{F}\to\mathcal{G}} = \|(\Lambda_j)_{j\in\mathbb{N}_0}\|_{\rho'} < \infty. \tag{7.34}$$

If f belongs to the $\mathcal{C}_{\boldsymbol{\lambda},\boldsymbol{n},a,b}$ defined in (7.32) and $n_0 = 0$, then

$$\begin{aligned}\|f\|_{\mathcal{F}} &= \left\|\left(\left\|\left(\frac{\widehat{f}(\boldsymbol{k})}{\lambda_{\boldsymbol{k}}}\right)_{\boldsymbol{k}\in\mathcal{K}_j}\right\|_\rho\right)_{j\in\mathbb{N}}\right\|_\rho = \|(\sigma_j(f))_{j\in\mathbb{N}}\|_\rho \\ &\le \|(\sigma_1(f), ab\sigma_1(f), ab^2\sigma_1(f), \dots)\|_\rho \\ &= \left(1 + \frac{a^\rho b^\rho}{1-b^\rho}\right)^{1/\rho}\left\|\left(\frac{\widehat{f}(\boldsymbol{k})}{\lambda_{\boldsymbol{k}}}\right)_{\boldsymbol{k}\in\mathcal{K}_1}\right\|_\rho.\end{aligned}$$

Comparing this inequality to the definition of $\mathcal{C}_{\boldsymbol{\lambda},n_1,A}$ in the previous section, it can be seen that $\mathcal{C}_{\boldsymbol{\lambda},\boldsymbol{n},a,b}$ (defined in (7.32)) is a subset of $\mathcal{C}_{\boldsymbol{\lambda},n_1,A}$ (defined in (7.18)) if we choose $A = [1 + a^\rho b^\rho/(1-b^\rho)]^{1/\rho}$ in (7.18).

From the expression for the error in (7.12) and the definition of the cone in (7.32), we can now derive a data-driven error bound for all $f \in \mathcal{C}_{\boldsymbol{\lambda},\boldsymbol{n},a,b}$ and $j \in \mathbb{N}$:

$$\begin{aligned}&\|\mathrm{SOL}(f) - \mathrm{APP}(f, n_j)\|_{\mathcal{G}} \\ &\quad = \|(\widehat{f}(\boldsymbol{k}_i))_{i=n_j+1}^{\infty}\|_\tau = \|(\|(\widehat{f}(\boldsymbol{k}))_{\boldsymbol{k}\in\mathcal{K}_l}\|_\tau)_{l=j+1}^{\infty}\|_\tau \\ &\quad \le \|(\sigma_l(f)\Lambda_l)_{l=j+1}^{\infty}\|_\tau \quad \text{by (7.33)} \\ &\quad = \|(\sigma_{j+l}(f)\Lambda_{j+l})_{l=1}^{\infty}\|_\tau \\ &\quad \le a\sigma_j(f)\|(b^l\Lambda_{j+l})_{l=1}^{\infty}\|_\tau =: \mathrm{ERR}((\widehat{f}(\boldsymbol{k}_i))_{i=1}^{n_j}, n_j) \quad \text{by (7.32).}\end{aligned} \tag{7.35}$$

This upper bound depends only on the function data and the parameters defining $\mathcal{C}$. The error vanishes as $j \to \infty$, because $\sigma_j(f) \le ab^{j-1}\sigma_1(f) \to 0$ and $\Lambda_j \to 0$. Moreover, the error bound for APP(f, n_j) depends on $\sigma_j(f)$, whose rate of decay need not be postulated in advance.

These assumptions accommodate both the cases, where the approximation converges algebraically and exponentially. To illustrate the algebraic case, suppose that

$\widehat{f}(\boldsymbol{k}_i)/\lambda_{\boldsymbol{k}_i} = \Theta(i^{-r_\Delta})$ for some positive $r_\Delta > 1/\rho$. Here, r_Δ reflects the smoothness of the input function. For this algebraic case, one would normally define $\mathcal{C}_{\boldsymbol{\lambda},\boldsymbol{n},a,b}$ in terms of an exponentially increasing sequence, $(n_j)_{j\geq 0}$, e. g., $n_j = n_0 2^j$, which implies that

$$\begin{aligned}\sigma_j(f) &= \left[\sum_{i=n_0 2^{j-1}+1}^{n_0 2^j} \left|\frac{\widehat{f}(\boldsymbol{k}_i)}{\lambda_{\boldsymbol{k}_i}}\right|^\rho\right]^{1/\rho} = \Theta\left(\left[\sum_{i=n_0 2^{j-1}+1}^{n_0 2^j} i^{-\rho r_\Delta}\right]^{1/\rho}\right)\\ &= \Theta(2^{-j(r_\Delta-1/\rho)}).\end{aligned}$$

Reasonable functions would satisfy

$$C_{\mathrm{lo}} 2^{-j(r_\Delta-1/\rho)} \leq \sigma_j(f) \leq C_{\mathrm{up}} 2^{-j(r_\Delta-1/\rho)}$$

for some constants C_{lo} and C_{up}. By reasonable, we mean that the values of $\sigma_j(f)$ would not drop precipitously and then rise again as j increases. Choosing $a \geq C_{\mathrm{up}}/C_{\mathrm{lo}}$ and $b \geq 2^{-(r_\Delta-1/\rho)}$ causes the cone $\mathcal{C}_{\boldsymbol{\lambda},\boldsymbol{n},a,b}$ to include such functions. Note that only the ratio of C_{up} to C_{lo} need be assumed to determine a, and choosing b larger than necessary does not affect the order of the decay of the error bound.

To illustrate the exponential case, suppose that $\widehat{f}(\boldsymbol{k}_i)/\lambda_{\boldsymbol{k}_i} = \Theta(\mathrm{e}^{-r_\Delta i})$. For this exponential case, one would normally define $\mathcal{C}$ in terms of an arithmetic sequence, $(n_j)_{j\geq 0}$, e. g., $n_j = n_0 + jq$, where q is a positive integer. This implies that

$$\begin{aligned}\sigma_j(f) &= \left[\sum_{i=n_0+jq-q+1}^{n_0+jq} \left|\frac{\widehat{f}(\boldsymbol{k}_i)}{\lambda_{\boldsymbol{k}_i}}\right|^\rho\right]^{1/\rho} = \Theta\left(\left[\sum_{i=n_0+jq-q+1}^{n_0+jq} \mathrm{e}^{-\rho r_\Delta i}\right]^{1/\rho}\right)\\ &= \Theta(\mathrm{e}^{-jr_\Delta q}).\end{aligned}$$

Analogous to the algebraic case, reasonable functions would satisfy $C_{\mathrm{lo}}\mathrm{e}^{-jr_\Delta q} \leq \sigma_j(f) \leq C_{\mathrm{up}}\mathrm{e}^{-jr_\Delta q}$ for some constants C_{lo} and C_{up}. Choosing $a \geq C_{\mathrm{up}}/C_{\mathrm{lo}}$ and $b \geq \mathrm{e}^{-r_\Delta q}$ causes the cone $\mathcal{C}_{\boldsymbol{\lambda},\boldsymbol{n},a,b}$ to include such functions. Again, only the ratio of C_{up} to C_{lo} need be assumed to determine a, and choosing b larger than necessary does not affect the order of the decay of the error bound.

7.4.1 The adaptive algorithm and its information cost

The data-driven error bound in (7.35) forms the basis for an adaptive Algorithm 7.3, which solves our problem for input functions in the cone $\mathcal{C}_{\boldsymbol{\lambda},\boldsymbol{n},a,b}$ defined in (7.32). The following theorem establishes its viability and information cost. In deriving upper bounds on the information cost and lower bounds on the complexity, we may sacrifice tightness for simplicity.

Algorithm 7.3 Adaptive ALG for a cone of input functions tracking the series coefficient decay rate.

Parameters: the Banach spaces $\mathcal{F}$ and $\mathcal{G}$, including the weights $\boldsymbol{\lambda}$; a strictly increasing sequence of nonnegative integers, $(n_j)_{j\geq 0}$; an inflation factor, a; the general decay rate, b; APP satisfying (7.12)
Input: a black-box function, f; an absolute error tolerance, $\varepsilon > 0$
Ensure: Error criterion (7.1) for the cone defined in (7.32)
Let $j \leftarrow 0$
repeat
Let $j \leftarrow j + 1$
Compute $\mathrm{ERR}((\widehat{f}(\boldsymbol{k}_i))_{i=1}^{n_j}, n_j)$ as defined in (7.35)
until $\mathrm{ERR}((\widehat{f}(\boldsymbol{k}_i))_{i=1}^{n_j}, n_j) \leq \varepsilon$
Return $\mathrm{ALG}(f,\varepsilon) = \mathrm{APP}(f, n_j)$

Theorem 7.14. *Algorithm 7.3 yields an answer satisfying absolute error criterion* (7.1), *i. e.,* $\mathrm{ALG} \in \mathcal{A}(\mathcal{C}_{\boldsymbol{\lambda},\boldsymbol{n},a,b})$ *for* $\mathcal{C}_{\boldsymbol{\lambda},\boldsymbol{n},a,b}$, *defined in* (7.32). *The information cost is* $\mathrm{COST}(\mathrm{ALG}, f, \varepsilon) = n_{j^*}$, *where* j^* *is defined implicitly as*

$$j^* = \min\{j \in \mathbb{N} : \mathrm{ERR}((\widehat{f}(\boldsymbol{k}_i))_{i=1}^{n_j}, n_j) \leq \varepsilon\}. \tag{7.36}$$

Moreover, $\mathrm{COST}(\mathrm{ALG}, \mathcal{C}_{\boldsymbol{\lambda},\boldsymbol{n},a,b}, \varepsilon, R) \leq n_{j^\dagger}$, *where* $j^\dagger$ *is defined as follows:*

$$j^\dagger = \min\left\{j \in \mathbb{N} : \|(b^l \Lambda_{j+l})_{l=1}^{\infty}\|_\tau \leq \frac{b\varepsilon}{Ra^2}\right\}. \tag{7.37}$$

Proof. The value of j^* in (7.36) follows directly from the error criterion. The success of the algorithm follows from the error bound in (7.35).

For the remainder of the proof, consider R and ε to be fixed. For any $f \in \mathcal{C}_{\boldsymbol{\lambda},\boldsymbol{n},a,b} \cap \mathcal{B}_R$, and for any $j^\dagger$ (defined as in (7.37)), it follows that

$$\begin{aligned}
R \geq \|f\|_{\mathcal{F}} &= \left\| \left(\frac{\widehat{f}(\boldsymbol{k})}{\lambda_{\boldsymbol{k}}} \right)_{\boldsymbol{k}\in\mathbb{K}} \right\|_\rho \geq \|(\sigma_j(f))_{j=1}^{j^\dagger}\|_\rho \quad \text{by (7.31)} \\
&\geq \sigma_1(f) \geq a^{-1} b^{1-j^\dagger} \sigma_{j^\dagger}(f) \quad \text{by (7.32)} \\
&\geq \sigma_{j^\dagger}(f) \frac{Ra}{\varepsilon} \|(b^l \Lambda_{j^\dagger+l})_{l=1}^{\infty}\|_\tau \quad \text{by the definition of } j^\dagger \text{ in (7.37)} \\
&= \frac{R}{\varepsilon} \mathrm{ERR}((\widehat{f}(\boldsymbol{k}_i))_{i=1}^{n_{j^\dagger}}, n_{j^\dagger}).
\end{aligned}$$

From this last inequality, it follows that $j^\dagger \geq j^*$. □

Although Algorithm 7.3 tracks the decay rate of the $\widehat{f}(\boldsymbol{k}_i)$, the information cost bound and complexity bound in the theorem above do not reflect different decay rates of the $\widehat{f}(\boldsymbol{k}_i)$. That is a subject for future investigation.

7.4.2 Bounds on the information cost for $\rho = \tau, \rho' = \infty$

We again turn to the special case considered in Section 7.3.2, where $\rho = \tau$ and $\rho' = \infty$. Furthermore, we consider a geometric progression of the n_j, namely, $\boldsymbol{n} = (n_0 2^j)_{j \in \mathbb{N}_0}$, so that

$$\mathcal{K}_0 = \{\boldsymbol{k}_1, \ldots, \boldsymbol{k}_{n_0}\}, \quad \mathcal{K}_j = \{\boldsymbol{k}_{n_0 2^{j-1}+1}, \ldots, \boldsymbol{k}_{n_0 2^j}\} \quad \text{for } j \in \mathbb{N}.$$

Our goal, as in Section 7.3.2, is to derive upper and lower bounds on the information cost of Algorithm 7.3. For the illustrative example in Section 7.2.2 with weights defined in (7.10), the sets $\mathcal{K}_j$ look roughly like hyperbolic slices of $\mathbb{N}_0^d$.

From Lemma 7.7, it follows that

$$\Lambda_j = \|(\lambda_{\boldsymbol{k}})_{\boldsymbol{k} \in \mathcal{K}_j}\|_\infty = \lambda_{\boldsymbol{k}_{n_0 2^{j-1}+1}} \le \frac{\|\boldsymbol{\lambda}\|_p}{n_0^{1/p} 2^{(j-1)/p}} \quad \forall j \in \mathbb{N}.$$

The key term in the upper bound on the information cost in (7.37) is

$$\begin{aligned}
\|(b^l \Lambda_{j+l})_{l=1}^\infty\|_\rho^\rho &= \sum_{l=1}^\infty (b^l \Lambda_{j+l})^\rho \\
&\le \sum_{l=1}^\infty \frac{b^{l\rho} \|\boldsymbol{\lambda}\|_p^\rho}{n_0^{\rho/p} 2^{(j+l-1)\rho/p}} = \frac{b^\rho \|\boldsymbol{\lambda}\|_p^\rho}{n_0^{\rho/p} 2^{j\rho/p}(1 - b^\rho/2^{\rho/p})}, \\
\|(b^l \Lambda_{j+l})_{l=1}^\infty\|_\rho &\le \frac{b\|\boldsymbol{\lambda}\|_p}{n_0^{1/p} 2^{j/p}(1 - b^\rho/2^{\rho/p})^{1/\rho}}.
\end{aligned}$$

From (7.37) in Theorem 7.14, it follows that the cost of Algorithm 7.3 has an upper bound of $\mathrm{COST}(\mathrm{ALG}, \mathcal{C}_{\boldsymbol{\lambda},\boldsymbol{n},a,b}, \varepsilon, R) \le n_j = n_0 2^j$ for any j satisfying

$$\begin{aligned}
\frac{b\|\boldsymbol{\lambda}\|_p}{n_0^{1/p} 2^{j/p}(1 - b^\rho/2^{\rho/p})^{1/\rho}} &\le \frac{b\varepsilon}{Ra^2}, \\
\left(\frac{Ra^2 \|\boldsymbol{\lambda}\|_p}{\varepsilon(1 - b^\rho/2^{\rho/p})^{1/\rho}}\right)^p &\le n_0 2^j.
\end{aligned}$$

This argument leads to the following upper bound on the information cost, which is similar to Theorem 7.8. Applying Lemma 7.9 yields a result analogous to Corollary 7.10.

Theorem 7.15. *If the elements of $\boldsymbol{\lambda}$ decay fast enough so that $\|\boldsymbol{\lambda}\|_p < \infty$ and $\rho' = \infty$, then the information cost of Algorithm 7.3 is bounded above by*

$$\begin{aligned}
&\mathrm{COST}(\mathrm{ALG}, \mathcal{C}_{\boldsymbol{\lambda},\boldsymbol{n},a,b}, \varepsilon, R) \\
&\quad \le \min\left\{n_0 2^j,\ j \in \mathbb{N}_0 : n_0 2^j \ge \left(\frac{Ra^2 \|\boldsymbol{\lambda}\|_p}{\varepsilon(1 - b^\rho/2^{\rho/p})^{1/\rho}}\right)^p\right\}.
\end{aligned}$$

For the function spaces defined in the illustrative example in Section 7.2.2, for all $p > 1/r$:

$$\operatorname{COST}(\mathrm{ALG}, \mathcal{C}_{\boldsymbol{\lambda},\boldsymbol{n},a,b}, \varepsilon, R) \le \min\left\{n_0 2^j,\ j \in \mathbb{N}_0 : n_0 2^j \ge \left(\frac{Ra^2 \exp(\zeta(pr)\|\boldsymbol{w}\|_p^p)}{\varepsilon(1 - b^\rho/2^{\rho/p})^{1/\rho}}\right)^p\right\}.$$

Thus, the information cost in this case is nearly $\mathcal{O}((R/\varepsilon)^{1/r})$, with a larger r yielding a lower asymptotic information cost.

7.4.3 Essential optimality of the algorithm

To establish the essential optimality of Algorithm 7.3 requires some additional, reasonable assumptions on the sequences $(n_j)_{j\in\mathbb{N}_0}$ and $(\sigma_j(f))_{j\in\mathbb{N}_0}$. Recall from (7.34) that $(\Lambda_j)_{j\in\mathbb{N}_0}$ has a finite ρ' norm. We require that the Λ_j must decay steadily with j:

$$\alpha^{-1}\beta^r \Lambda_j \le \Lambda_{j+r} \le \alpha\gamma^r \Lambda_j \quad \forall j, r \in \mathbb{N}_0, \quad \text{for some } \beta, \gamma < 1 \le \alpha. \tag{7.38}$$

We also assume that the ratio of the largest to smallest $\lambda_{\boldsymbol{k}}$ in a group is bounded above:

$$\sup_{j\in\mathbb{N}} \frac{\lambda_{\boldsymbol{k}_{n_{j-1}+1}}}{\lambda_{\boldsymbol{k}_{n_j}}} \le S_1 < \infty. \tag{7.39}$$

For the illustrative choices of $(n_j)_{j\in\mathbb{N}_0}$ and $(\boldsymbol{k}_i)_{i\in\mathbb{N}}$ preceding Section 7.4.1, this assumption holds. Let card(·) denote the cardinality of a set. We assume that if $\mathcal{J}$ is an arbitrary set of wavenumbers with $\operatorname{card}(\mathcal{J}) \le n_j$, then there exists some $l \le j+1$, for which $\mathcal{K}_l \setminus \mathcal{J}$ retains some significant fraction of the original $\mathcal{K}_l$ elements:

$$\inf_{j\in\mathbb{N}} \min_{\mathcal{J}\subset\mathbb{K}:\operatorname{card}(\mathcal{J})\le n_j} \max_{0\le l\le j+1} \frac{\operatorname{card}(\mathcal{K}_l \setminus \mathcal{J})}{\operatorname{card}(\mathcal{K}_l)} \ge S_2 > 0. \tag{7.40}$$

Again, for the illustrative choices of $(n_j)_{j\in\mathbb{N}_0}$ and $(\boldsymbol{k}_i)_{i\in\mathbb{N}}$ preceding Section 7.4.1, this assumption holds.

The following theorem establishes a lower bound on the complexity of our problem for input functions in $\mathcal{C}_{\boldsymbol{\lambda},\boldsymbol{n},a,b}$. The theorem after that shows that the cost of our algorithm as given in Theorem 7.14 is essentially no worse than this lower bound.

Theorem 7.16. *A lower bound on the complexity of the linear problem is*

$$\operatorname{COMP}(\mathcal{C}_{\boldsymbol{\lambda},\boldsymbol{n},a,b}, \varepsilon, R) > n_{j^\ddagger},$$

where

$$j^\ddagger = \max\left\{j \in \mathbb{N} : b^{j+1}\Lambda_{j+1} > \frac{2a\alpha\varepsilon}{R(a-1)(1-b^\rho)^{1/\rho}}\left[1 + \left(\frac{1}{S_2} - 1\right)S_1^\rho\right]^{1/\rho}\right\}.$$

Proof. As in the proof of Theorem 7.5, we consider fixed and arbitrary R and ε. We proceed by carefully constructing the test input functions, f_1 and $f_\pm = f_1 \pm f_2$, lying in $\mathcal{C}_{\boldsymbol{\lambda},\boldsymbol{n},a,b} \cap \mathcal{B}_R$, which yield the same approximate solution, but different true solutions. This leads to a lower bound on $\operatorname{COMP}(\mathcal{C}_{\boldsymbol{\lambda},\boldsymbol{n},a,b}, \varepsilon, R)$. The proof is provided for $\rho' < \infty$. The proof for $\rho' = \infty$ is similar.

The first test function $f_1 \in \mathcal{C}_{\boldsymbol{\lambda},\boldsymbol{n},a,b}$ is defined in terms of its series coefficients—inspired by Lemma 7.1—as

$$f_1 = f_{10} + f_{11} + \cdots, \quad \widehat{f}_{1j}(\boldsymbol{k}) := \begin{cases} \dfrac{c_1 b^j \lambda_{\boldsymbol{k}}^{\rho'/\rho+1}}{\Lambda_j^{\rho'/\rho}}, & \boldsymbol{k} \in \mathcal{K}_j, \\ 0, & \boldsymbol{k} \notin \mathcal{K}_j, \end{cases}$$

$$c_1 := \frac{R(a+1)(1-b^\rho)^{1/\rho}}{2a}.$$

It can be verified that the test function lies both in $\mathcal{B}_R$ and in $\mathcal{C}_{\boldsymbol{\lambda},\boldsymbol{n},a,b}$:

$$\begin{aligned} \sigma_j(f_1) &= \left\| \left(\frac{\widehat{f}_{1j}(\boldsymbol{k})}{\lambda_{\boldsymbol{k}}} \right)_{\boldsymbol{k} \in \mathcal{K}_j} \right\|_\rho = c_1 b^j, \quad j \in \mathbb{N}_0, \\ \|f_1\|_{\mathcal{F}} &= \|(\sigma_j(f))_{j \in \mathbb{N}_0}\|_\rho = \frac{c_1}{(1-b^\rho)^{1/\rho}} = \frac{R(a+1)}{2a} \le R, \\ \sigma_{j+r}(f_1) &= b^l \sigma_j(f_1) \le a b^l \sigma_j(f_1), \quad j, l \in \mathbb{N}_0. \end{aligned}$$

Now, let ALG' be an arbitrary algorithm in $\mathcal{A}(\mathcal{C}_{\boldsymbol{\lambda},\boldsymbol{n},a,b})$, and suppose that $\mathrm{ALG}'(f_1, \varepsilon)$ samples $f_1(\boldsymbol{k})$ for $\boldsymbol{k} \in \mathcal{J}$. Let $\widetilde{\mathcal{K}}_j = \mathcal{K}_j \setminus \mathcal{J}$ for all nonnegative integers j. Construct the function f_2, having zero coefficients for $\boldsymbol{k} \in \mathcal{J}$, but otherwise looking like f_1:

$$f_2 = f_{20} + f_{21} + \cdots, \quad \widehat{f}_{2j}(\boldsymbol{k}) := \begin{cases} \dfrac{c_2 b^j \lambda_{\boldsymbol{k}}^{\rho'/\rho+1}}{\Lambda_j^{\rho'/\rho}}, & \boldsymbol{k} \in \widetilde{\mathcal{K}}_j, \\ 0, & \text{otherwise}, \end{cases}$$

$$\begin{aligned} c_2 &:= \frac{R(a-1)(1-b^\rho)^{1/\rho}}{2a}, \quad \widetilde{\Lambda}_j := \|(\lambda_{\boldsymbol{k}})_{\boldsymbol{k} \in \widetilde{\mathcal{K}}_j}\|_{\rho'} \le \Lambda_j, \\ \sigma_j(f_2) &= \left\| \left(\frac{\widehat{f}_{2j}(\boldsymbol{k})}{\lambda_{\boldsymbol{k}}} \right)_{\boldsymbol{k} \in \widetilde{\mathcal{K}}_j} \right\|_\rho = \begin{cases} c_2 b^j, & \widetilde{\mathcal{K}}_j \ne \emptyset, \\ 0, & \widetilde{\mathcal{K}}_j = \emptyset, \end{cases} \quad j \in \mathbb{N}_0, \\ \|f_2\|_{\mathcal{F}} &= \|(\sigma_j(f_2))_{j \in \mathbb{N}_0}\|_\rho \le \frac{c_2}{(1-b^\rho)^{1/\rho}} = \frac{R(a-1)}{2a} \le R, \\ \|\mathrm{SOL}(f_{2j})\|_{\mathcal{G}} &= \sigma_j(f_{2j}) \widetilde{\Lambda}_j = c_2 b^j \widetilde{\Lambda}_j, \quad j \in \mathbb{N}_0, \\ \|\mathrm{SOL}(f_2)\|_{\mathcal{G}} &= \|(c_2 b^j \widetilde{\Lambda}_j)_{j \in \mathbb{N}_0}\|_\tau = \frac{R(a-1)(1-b^\rho)^{1/\rho}}{2a} \|(b^j \widetilde{\Lambda}_j)_{j \in \mathbb{N}_0}\|_\tau. \end{aligned} \tag{7.41}$$

Furthermore, define $f_\pm = f_1 \pm f_2$. It can be verified that $f_\pm$ also lie both in $\mathcal{B}_R$ and in $\mathcal{C}_{\boldsymbol{\lambda},\boldsymbol{n},a,b}$:

$$\begin{aligned}
\|f_\pm\|_{\mathcal{F}} &\le \|f_1\|_{\mathcal{F}} + \|f_2\|_{\mathcal{F}} \le \frac{c_1 + c_2}{(1-b^\rho)^{1/\rho}} = R, \\
\sigma_j(f_\pm) &\ge \sigma_j(f_1) - \sigma_j(f_2) \ge (c_1 - c_2)b^j = \frac{R(1-b^\rho)^{1/\rho}b^j}{a}, \quad j \in \mathbb{N}_0, \\
\sigma_{j+l}(f_\pm) &\le \sigma_{j+l}(f_1) + \sigma_{j+l}(f_2) \le (c_1 + c_2)b^{j+l} \\
&= R(1-b^\rho)^{1/\rho} b^{j+l} \le ab^l \sigma_j(f_\pm), \quad j, l \in \mathbb{N}_0.
\end{aligned}$$

Since $\widehat{f}_2(\boldsymbol{k}) = 0$ for $\boldsymbol{k} \in \mathcal{J}$, it follows that $\mathrm{ALG}'(f_\pm, \varepsilon) = \mathrm{ALG}'(f_1, \varepsilon)$. But, though the two test functions $f_\pm$ lead to the same approximate solution, they have different true solutions. In particular,

$$\begin{aligned}
\varepsilon &\ge \max\{\|\mathrm{SOL}(f_+) - \mathrm{ALG}'(f_+, \varepsilon)\|_{\mathcal{G}}, \|\mathrm{SOL}(f_-) - \mathrm{ALG}'(f_-, \varepsilon)\|_{\mathcal{G}}\} \\
&\ge \frac{1}{2}[\|\mathrm{SOL}(f_+) - \mathrm{ALG}(f_1, \varepsilon)\|_{\mathcal{G}} + \|\mathrm{SOL}(f_-) - \mathrm{ALG}'(f_1, \varepsilon)\|_{\mathcal{G}}] \\
&\quad \text{since } \mathrm{ALG}'(f_\pm, \varepsilon) = \mathrm{ALG}'(f_1, \varepsilon) \\
&\ge \frac{1}{2}\|\mathrm{SOL}(f_+) - \mathrm{SOL}(f_-)\|_{\mathcal{G}} \quad \text{by the triangle inequality} \\
&\ge \frac{1}{2}\|\mathrm{SOL}(f_+ - f_-)\|_{\mathcal{G}} \quad \text{since SOL is linear} \\
&= \|\mathrm{SOL}(f_2)\|_{\mathcal{G}} = \frac{R(a-1)(1-b^\rho)^{1/\rho}}{2a} \|(b^j \widetilde{\Lambda}_j)_{j\in\mathbb{N}_0}\|_\tau \quad \text{by (7.41).}
\end{aligned} \tag{7.42}$$

Suppose that $\mathrm{card}(\mathcal{J}) = \mathrm{COST}(\mathrm{ALG}', f_\pm, \varepsilon) \le n_{j^\star}$. Then by condition (7.40), there exists an $l^\star \le j^\star + 1$, where $\mathrm{card}(\widetilde{\mathcal{K}}_{l^\star}) \ge S_2 \,\mathrm{card}(\mathcal{K}_{l^\star})$. This implies a lower bound on $\Lambda_{l^\star}$. Let $m = n_{l^\star} - n_{l^\star - 1} = \mathrm{card}(\mathcal{K}_{l^\star})$. Then, $m_{\mathrm{in}} = \lceil S_2 m \rceil \ge S_2 m$ is a lower bound on $\mathrm{card}(\widetilde{\mathcal{K}}_{l^\star})$, and $m_{\mathrm{out}} = m - m_{\mathrm{in}} \le (1 - S_2)m$ is an upper bound on $\mathrm{card}(\mathcal{K}_{l^\star} \setminus \widetilde{\mathcal{K}}_{l^\star})$. Moreover,

$$\begin{aligned}
\Lambda_{l^\star}^\rho &= \sum_{i\in\mathcal{K}_{l^\star}} \lambda_{\boldsymbol{k}_i}^\rho = \widetilde{\Lambda}_{l^\star}^\rho + \sum_{i\in\mathcal{K}_{l^\star}\setminus\widetilde{\mathcal{K}}_{l^\star}} \lambda_{\boldsymbol{k}_i}^\rho \\
&\le \widetilde{\Lambda}_{l^\star}^\rho + m_{\mathrm{out}} \lambda_{\boldsymbol{k}_{n_{l^\star-1}+1}}^\rho \quad \text{by the ordering of the } \lambda_{\boldsymbol{k}_i} \\
&\le \widetilde{\Lambda}_{l^\star}^\rho + m_{\mathrm{out}} S_1^\rho \lambda_{\boldsymbol{k}_{n_{l^\star}}}^\rho \quad \text{by (7.39)} \\
&\le \widetilde{\Lambda}_{l^\star}^\rho + \frac{m_{\mathrm{out}}}{m_{\mathrm{in}}} S_1^\rho \Lambda_{l^\star}^\rho \quad \text{by the definition of } \widetilde{\Lambda}_{l^\star} \\
&\le \left[1 + \left(\frac{1}{S_2} - 1\right) S_1^\rho\right] \widetilde{\Lambda}_{l^\star}^\rho \quad \text{by the bounds on } m_{\mathrm{in}} \text{ and } m_{\mathrm{out}} \\
&\le \left[1 + \left(\frac{1}{S_2} - 1\right) S_1^\rho\right] b^{-\rho l^\star} \|(b^j \widetilde{\Lambda}_j)_{j\in\mathbb{N}_0}\|_\tau^\rho .
\end{aligned}$$

Returning to (7.42), the above inequality implies that

$$\varepsilon \geq \frac{R(a-1)(1-b^{\rho})^{1/\rho}}{2a}\left[1+\left(\frac{1}{S_2}-1\right)S_1^{\rho}\right]^{-1/\rho} b^{l^*}\Lambda_{l^*}.$$

Since $l^* \leq j^*+1$, it follows that $b^{l^*} \geq b^{j^*+1}$; from condition (7.38), it follows that $\Lambda_{l^*} \geq \Lambda_{j^*+1}/\alpha$. Thus,

$$\varepsilon \geq \frac{R(a-1)(1-b^{\rho})^{1/\rho}}{2a\alpha}\left[1+\left(\frac{1}{S_2}-1\right)S_1^{\rho}\right]^{-1/\rho} b^{j^*+1}\Lambda_{j^*+1}.$$

If any algorithm satisfies the error tolerance ε for all input functions in $\mathcal{C}_{\Lambda,n,a,b} \cap \mathcal{B}_R$ and has information cost no greater than n_{j^*}, then j^* must satisfy the above inequality. By contrast, if the above inequality is violated for any j^*, then the information cost of the successful algorithm must be greater than n_{j^*}. This completes the proof. □

Theorem 7.17. *Adaptive Algorithm 7.3 is essentially optimal for the cone of input functions defined in (7.32).*

Proof. Let $j^\dagger(\varepsilon)$ be defined as in (7.37), with the ε dependence made explicit. Choose $\varepsilon_{\max}$ and $R_{\min}$ in (7.2) such that $j^\dagger(\varepsilon) \geq 2$. This definition implies that

$$\begin{aligned}
b^{j^\dagger(\varepsilon)}\Lambda_{j^\dagger(\varepsilon)} &= \frac{[1-(\gamma b)^{\tau}]^{1/\tau}}{\alpha}\left\|\left(b^{j^\dagger(\varepsilon)-1+r}\alpha\gamma^{r-1}\Lambda_{j^\dagger(\varepsilon)}\right)_{r=1}^{\infty}\right\|_{\tau} \\
&\geq \frac{[1-(\gamma b)^{\tau}]^{1/\tau}}{\alpha}\left\|\left(b^{j^\dagger(\varepsilon)-1+r}\Lambda_{j^\dagger(\varepsilon)-1+r}\right)_{r=1}^{\infty}\right\|_{\tau} \quad \text{by (7.38)} \\
&> \frac{b[1-(\gamma b)^{\tau}]^{1/\tau}\varepsilon}{Ra^2\alpha}\left(\frac{1-b^{\rho(j^\dagger(\varepsilon)-1)}}{1-b^{\rho}}\right)^{1/\rho} \quad \text{by (7.37)} \\
&\geq \frac{b[1-(\gamma b)^{\tau}]^{1/\tau}\varepsilon}{Ra^2\alpha} \quad \text{since } j^\dagger(\varepsilon) \geq 2 \\
&= \frac{\alpha^2}{(b\beta)^2} \times \frac{2a\alpha\omega\varepsilon}{R(a-1)(1-b^{\rho})^{1/\rho}}\left[1+\left(\frac{1}{S_2}-1\right)S_1^{\rho}\right]^{1/\rho},
\end{aligned}$$

where

$$\omega = \frac{(a-1)b^3\beta^2(1-b^{\rho})^{1/\rho}[1-(\gamma b)^{\tau}]^{1/\tau}}{2a^3\alpha^4}\left[1+\left(\frac{1}{S_2}-1\right)S_1^{\rho}\right]^{-1/\rho}.$$

Making the ε dependence explicit in the definition of $j^\ddagger(\varepsilon)$ in Theorem 7.16, it follows from the above inequality and (7.38) that

$$b^{j^\dagger(\varepsilon)}\Lambda_{j^\dagger(\varepsilon)} > \frac{\alpha^2}{(b\beta)^2}b^{j^\ddagger(\omega\varepsilon)+2}\Lambda_{j^\ddagger(\omega\varepsilon)+2} \geq \alpha b^{j^\ddagger(\omega\varepsilon)}\Lambda_{j^\ddagger(\omega\varepsilon)}.$$

If $j^\dagger(\varepsilon) \geq j^\ddagger(\omega\varepsilon)$, then (7.38) implies that

$$b^{j^\dagger(\varepsilon)}\Lambda_{j^\dagger(\varepsilon)} \leq \alpha(\gamma b)^{j^\dagger(\varepsilon)-j^\ddagger(\omega\varepsilon)}b^{j^\ddagger(\omega\varepsilon)}\Lambda_{j^\ddagger(\omega\varepsilon)} \leq \alpha b^{j^\ddagger(\omega\varepsilon)}\Lambda_{j^\ddagger(\omega\varepsilon)}.$$

But, this contradicts the above inequality. Thus, $j^\dagger(\varepsilon) < j^\ddagger(\omega\varepsilon)$, and so

$$\mathrm{COST}(\mathrm{ALG}, \mathcal{C}_{\boldsymbol{\lambda},\boldsymbol{n},a,b}, \varepsilon, R) \le n_{j^\dagger(\varepsilon)} < n_{j^\ddagger(\omega\varepsilon)} < \mathrm{COMP}(\mathcal{C}_{\boldsymbol{\lambda},\boldsymbol{n},a,b}, \omega\varepsilon, R).$$

Thus, Algorithm 7.3 is essentially optimal. □

7.4.4 Tractability

We again would like to study tractability. As it turns out, by using the relation between the cones defined in (7.18) and (7.32), respectively, we easily obtain sufficient conditions for tractability.

Theorem 7.18. *The respective conditions presented in Theorem 7.11 for the case where $\rho' = \infty$ and in Theorem 7.12 for the case where $\rho' < \infty$ are sufficient for strong polynomial, polynomial, and weak tractability of the approximation problem defined on cones as in* (7.32).

Proof. As pointed out above, $\mathcal{C}_{\boldsymbol{\lambda},\boldsymbol{n},a,b}$ defined in (7.32) is a subset of $\mathcal{C}_{\boldsymbol{\lambda},n_1,A}$ defined in (7.18), by choosing $A = (1 + \frac{a^\rho b^\rho}{1-b^\rho})^{1/\rho}$ in (7.18). This means that the approximation problem on $\mathcal{C}_{\boldsymbol{\lambda},\boldsymbol{n},a,b}$ defined in (7.32) is essentially (i. e., up to constants depending on A, a, b and ρ) no harder than the same problem on $\mathcal{C}_{\boldsymbol{\lambda},\boldsymbol{n},a,b}$ defined in (7.18). This, however, implies that all sufficient conditions in Theorem 7.11 are also sufficient in the case considered in Theorem 7.18. □

Theorem 7.18 yields sufficient conditions for the tractability notions considered here. A general result for necessary conditions seems to be more difficult to obtain and is left open for future research.

7.5 Inferring coordinate and smoothness importance

In Sections 7.3 and 7.4, the weights $\boldsymbol{\lambda} = (\lambda_{\boldsymbol{k}_1}, \lambda_{\boldsymbol{k}_2}, \ldots)$, which appear in the definition of the cone of inputs, $\mathcal{C}_{\boldsymbol{\lambda},n_1,A}$ and $\mathcal{C}_{\boldsymbol{\lambda},\boldsymbol{n},a,b}$, are taken as given and fixed. One may assume the form suggested in (7.25), which defines $\boldsymbol{\lambda}$ in terms of coordinate weights, $\boldsymbol{w}$, and smoothness weights, $\boldsymbol{s}$. However, practically speaking it may be difficult to know a priori suitable values for these weights. This section explores a situation, where the initial data collected for the input function data can be used to learn $\boldsymbol{w}$ and $\boldsymbol{s}$, inferring which input variables in f may be more important and the smoothness of the function. We focus on inferring the $\boldsymbol{\lambda}$ defining the cone $\mathcal{C}_{\boldsymbol{\lambda},n_1,A}$ in Section 7.3.

This section is motivated by situations, where the relative importance of the d input variables of the function is not known from physical considerations. We also envision situations, where the cost of function data is large, e. g., the result of an expensive computer simulation. Thus, we are not concerned with the cost of the algorithm beyond the information cost, which we hope to limit to $\mathcal{O}(d)$.

7.5.1 Product-, order- and smoothness-dependent (POSD) weights

The $u_{\boldsymbol{k}}$ and the $\lambda_{\boldsymbol{k}}$ in this section are defined as

$$u_{\boldsymbol{k}} = \prod_{\ell=1}^{d} u_{k_\ell}, \quad \lambda_{\boldsymbol{k}} = \Gamma_{\|\boldsymbol{k}\|_0} \prod_{\substack{\ell=1 \\ k_\ell>0}}^{d} w_\ell s_{k_\ell}, \quad \Gamma_0 = s_1 = 1, \quad \boldsymbol{k} \in \mathbb{N}_0^d, \tag{7.43}$$

where $\boldsymbol{w} = (w_\ell)_{\ell=1}^d$ is the vector of coordinate weights, $\boldsymbol{s} = (s_k)_{k=1}^\infty$ is the vector of smoothness weights, $\boldsymbol{\Gamma} = (\Gamma_m)_{m=1}^d$ is the vector of *order weights*, and $\|\boldsymbol{k}\|_0$ denotes the number of nonzero elements of $\boldsymbol{k}$. This is a generalization of the form for $\boldsymbol{\lambda}$ given by (7.25). The intuition behind these weights is as follows:

- Coordinate weights quantify the importance for the d input variables in f.
- Smoothness weights quantify the importance of the u_k. E. g., if the u_k are polynomials of degree k, as in Section 7.2.2, then the faster the s_k decay, the smoother f is.
- Order weights quantify the importance of effects with different orders; $\boldsymbol{k}$ having one nonzero element corresponds to a first-order or main effect; $\boldsymbol{k}$ having two nonzero elements corresponds to a second-order (interaction) effect (e. g., first-order, second-order).

This parameterization is motivated by several guiding principles from the experimental design literature [15], which are briefly described below. In statistical parlance, the terms $\widehat{f}(\boldsymbol{k})u_{\boldsymbol{k}}$ are effects.

- *Effect sparsity* assumes that only a small number of coordinates in f are important. In (7.43), this sparsity means that only a small number of product weights $\boldsymbol{w}$ are large. This principle arises in the sufficient condition for strong tractability in Corollary 7.13.
- *Effect heredity* assumes that lower-order effects are more important than higher-order effects. E. g., $\lambda_{(1,0,0,\dots,0)}$ should be larger than $\lambda_{(1,1,0,\dots,0)}$. In (7.43), this heredity can be enforced by assuming that the order weights Γ_m decrease with m.
- *Effect hierarchy* assumes that an effect is active *only* when all its component effects are active. For example, $\lambda_{(1,1,0,\dots,0)} > 0$ only when $\lambda_{(1,0,0,\dots,0)}$ and $\lambda_{(0,1,0,\dots,0)}$ are both nonzero. This hierarchy is implicitly enforced by the product structure of the weights in (7.43).

- *Effect smoothness* assumes that lower-degree effects are more important than higher-degree effects. For example, when the $(u_k)_{k\in\mathbb{N}_0}$ are polynomials, this means that linear effects are more important than quadratic effects, which are in turn more significant than cubic effects, and so on. Effect smoothness can be imposed by assuming $\boldsymbol{s}$ to be a decreasing sequence.

These principles will guide the inference of weights $\boldsymbol{\lambda}$ (and its corresponding cone of inputs $\mathcal{C}_{\boldsymbol{\lambda},n_1,A}$) from initial data.

The $\lambda_{\boldsymbol{k}}$ defined in (7.43) are called product-, order- and smoothness-dependent (POSD) weights. From a quasi-Monte Carlo (QMC) perspective, the POSD weights in (7.43) generalize upon the product-and-order dependent (POD) weights in [6], which were introduced for analyzing QMC methods in partial differential equations with random coefficients. The latter POD weights can be recovered by ignoring the smoothness weights.

Our POSD weights differ from the smoothness-driven product-and-order dependent (SPOD) weights in [1], which were recently used to analyze higher-order QMC methods for stochastic partial differential equations. These SPOD weights take the form:

$$\gamma_{\mathfrak{u}} = \sum_{\boldsymbol{k}\in\{1,\ldots,\alpha\}^{|\mathfrak{u}|}} \|\boldsymbol{k}\|_1! \prod_{\ell\in\mathfrak{u}} (2^{\delta(k_\ell,\alpha)} w_\ell^{k_\ell}), \quad \|\boldsymbol{k}\|_1 = \sum_{\ell=1}^{d} k_\ell, \quad \mathfrak{u} \subseteq \{1,\ldots,d\},$$

where $\delta(k_\ell,\alpha)$ is 1 if $k_\ell = \alpha$ and 0 otherwise. Intuitively, the SPOD weights quantify the importance of each *subspace* (indexed by $\mathfrak{u}$), under a common smoothness structure among subspaces (for further details on SPOD weights, we refer the reader to [1]). In contrast, the proposed POSD weights in (7.43) instead quantify the importance of each *Fourier series coefficient* $\widehat{f}(\boldsymbol{k})$ (indexed by $\boldsymbol{k}$), under a common smoothness structure among coefficients.

7.5.2 Inferring POSD weights from an initial sample

As mentioned above, our goal here is to infer $\boldsymbol{\lambda}$ from input function data. We start with an initial set of $dk_{\max} + 1$ wavenumbers:

$$\bar{\mathcal{K}} = \{(0,\ldots,0,k,0,\ldots,0) : k = 0,\ldots,k_{\max}\}, \tag{7.44}$$

where the possible nonzero entry k may be at any one position in the vector. The approximation to f based on sampling the series coefficients for these wavenumbers is

$$f_{\mathrm{app}} = \sum_{\boldsymbol{k}\in\bar{\mathcal{K}}} \widehat{f}(\boldsymbol{k}) u_{\boldsymbol{k}}.$$

To choose the cone $\mathcal{C}_{\boldsymbol{\lambda},n_1,A}$, which best fits f, we select the POSD weights $\boldsymbol{\lambda}$ to make the norm of f_{app} small. These weights can be written as:

$$\bar{\boldsymbol{\lambda}} = \boldsymbol{\lambda}(\bar{\boldsymbol{w}}, \bar{\boldsymbol{s}}, \boldsymbol{\Gamma}),$$
$$\text{where } (\bar{\boldsymbol{w}}, \bar{\boldsymbol{s}}) = \min\left\{ \underset{(\boldsymbol{w},\boldsymbol{s})\in\mathcal{W}\times\mathcal{S}}{\operatorname{argmin}} \left\| \left(\frac{\widehat{f}(\boldsymbol{k})}{\lambda_{\boldsymbol{k}}(\boldsymbol{w}, \boldsymbol{s}, \boldsymbol{\Gamma})} \right)_{\boldsymbol{k}\in\mathcal{K}} \right\|_p \right\}. \tag{7.45}$$

Here, $\mathcal{W}$ is a candidate set for coordinate weights, e. g., $\mathcal{W} = [0, w^*]^d$, and $\mathcal{S}$ is a candidate set for the smoothness weights, e. g., $\mathcal{S} = \{(1/k^r)_{k=1}^{\infty} : r > 0\}$. The inner minimization finds the $(\boldsymbol{w}, \boldsymbol{s})$ that minimizes the approximate norm of the input function. This minimizer may be nonunique, so the outer minimization chooses the smallest such $(\boldsymbol{w}, \boldsymbol{s})$. Making the coordinate and smoothness weights as small as possible helps enforce the principles of effect sparsity. The optimum, $(\bar{\boldsymbol{w}}, \bar{\boldsymbol{s}})$, then defines the *data-inferred* POSD $\boldsymbol{\lambda}$, denoted $\bar{\boldsymbol{\lambda}}$.

The candidate sets $\mathcal{W}$ and $\mathcal{S}$ should be constructed such that the coordinate and smoothness weights have a priori upper bounds. Otherwise the inner minimization would choose huge values for $\boldsymbol{w}$ and $\boldsymbol{s}$ to maximize the $\lambda_{\boldsymbol{k}}(\boldsymbol{w}, \boldsymbol{s}, \boldsymbol{\Gamma})$, and minimize the norm of f_{app}. The cardinality of the initial set of wavenumbers is $dk_{\max} + 1$. There is a trade-off between keeping $k_{\max}$ small enough to reduce cost, and making $k_{\max}$ large enough to ensure robustness.

For simplicity, we assume that order weights, $\boldsymbol{\Gamma}$, are fixed a priori. If desired, they too could be inferred as the next step. However, since we want to limit the size of the initial sample to $\mathcal{O}(d)$, we must sample judiciously the higher-order interactions.

The optimization in (7.45) is nontrivial to solve numerically. In practice, we iteratively optimize over $\boldsymbol{w}$, and then $\boldsymbol{s}$, until convergence is reached. At each step of the iteration $\|(\widehat{f}(\boldsymbol{k})/\lambda_{\boldsymbol{k}})_{\boldsymbol{k}\in\mathcal{K}}\|_p$ decreases.

Algorithm 7.4 combines the construction of data-inferred POSD weights, $\bar{\boldsymbol{\lambda}}$, with Algorithm 7.2 of Section 7.3. This algorithm succeeds for input functions in the cone

$$\bar{\mathcal{C}}_{n_1,A} := \{f \in \mathcal{F} : f \in \mathcal{C}_{\bar{\boldsymbol{\lambda}},n_1,A} \text{ for } \bar{\boldsymbol{\lambda}} \text{ defined in (7.45)}\}. \tag{7.46}$$

The reason that $\bar{\mathcal{C}}_{n_1,A}$ is a cone is that the data-inferred $\bar{\boldsymbol{\lambda}}$ for the input function f is exactly the same as for the input function cf, where c is any constant.

7.5.3 Numerical Examples

We now investigate the numerical performance of this adaptive algorithm using data-inferred POSD weights. For simplicity, only the case of $p = \infty$ and $p' = \tau = 1$ is considered in the following examples. Here, the basis functions $(u_{\boldsymbol{k}})_{\boldsymbol{k}\in\mathbb{N}_0^d}$ are the Chebyshev polynomials in Section 7.2.2, and the solution operator is $\text{SOL}(f) = f$ (i. e., function

Algorithm 7.4 Adaptive ALG based on data-inferred POSD weights.

Parameters: the bases $\{u_{\boldsymbol{k}}\}_{\boldsymbol{k}\in\mathbb{N}_0^d}$ and $\{v_{\boldsymbol{k}}\}_{\boldsymbol{k}\in\mathbb{N}_0^d}$; candidate sets $\mathcal{W}$ and $\mathcal{S}$ for product and smoothness weights; maximum smoothness degree, $k_{\max}$; order weights, $\boldsymbol{\Gamma}$; an inflation factor, $A > 1$; APP satisfying (7.13)

Input: a black-box function, f; an absolute error tolerance, $\varepsilon > 0$

Ensure: Error criterion (7.1) for the cone defined in (7.46)

Define the initial set of wavenumbers $\tilde{\mathcal{K}}$ defined in (7.44)

Evaluate initial sample $\{\hat{f}(\boldsymbol{k})\}_{\boldsymbol{k}\in\tilde{\mathcal{K}}}$

Compute data-driven POSD weight $\bar{\boldsymbol{\lambda}}$, by iteratively optimizing over $(\boldsymbol{w},\boldsymbol{s}) \in \mathcal{W}\times\mathcal{S}$ in (7.45) until convergence

Using these weights, $\bar{\boldsymbol{\lambda}}$, perform Algorithm 7.2 to obtain ALG(f,ε)

Return ALG(f,ε)

approximation). We note that $\|f - \mathrm{ALG}(f,\varepsilon)\|_\infty \le \|f - \mathrm{ALG}(f,\varepsilon)\|_{\mathcal{G}}$, so our error criterion (7.1) implies that $\|f - \mathrm{ALG}(f,\varepsilon)\|_\infty \le \varepsilon$.

The simulation set-up is as follows: The Fourier coefficients for input function f, $\{\hat{f}(\boldsymbol{k})\}_{\boldsymbol{k}\in\mathbb{N}_0^d}$, are randomly sampled as

$$\hat{f}(\boldsymbol{k}) = Z_{\boldsymbol{k}}\,\Gamma^{\mathrm{tr}}_{\|\boldsymbol{k}\|_0}\prod_{\substack{\ell=1\\ k_\ell>0}}^{d} w_\ell^{\mathrm{tr}} s_{k_\ell}^{\mathrm{tr}}, \quad Z_{\boldsymbol{k}} \overset{i.i.d.}{\sim} \mathrm{Unif}[-1,1], \quad \boldsymbol{k}\in\mathbb{N}_0^d.$$

Here, $(w_\ell^{\mathrm{tr}})_{\ell=1}^d = (1/L^2(\ell))_{\ell=1}^d$, $(\Gamma_m^{\mathrm{tr}})_{m=1}^\infty \equiv 1$, and $(s_k^{\mathrm{tr}})_{k=1}^{k_{\max}} = (1/k^4)_{k=1}^4$ are the true coordinate, order, and smoothness weights, and $Z_{\boldsymbol{k}}$ randomly sets the magnitude and sign of each coefficient. Moreover, $(L(\ell))_{\ell=1}^d$ is a random permutation of $1,\dots,d$ to ensure that the order of input variables does not necessarily reflect their order of importance. We also set $\boldsymbol{\Gamma} = \boldsymbol{\Gamma}^{\mathrm{tr}}$ in Algorithm 7.4 and use an inflation factor of $A = 1.1$.

Figures 7.1 (a) and (b) display the total required sample size from Algorithm 7.4, as a function of the error to tolerance ratio, $\|f - \mathrm{ALG}(f,\varepsilon)\|_\infty/\varepsilon$, in $d = 4$ and $d = 7$ dimensions, respectively. Each data point corresponds to a different error tolerance ε. A ratio $\|f - \mathrm{ALG}(f,\varepsilon)\|_\infty/\varepsilon$ close to, but not exceeding, one is desired, since this shows that our adaptive algorithm is successful. For $d = 4$, $\|f - \mathrm{ALG}(f,\varepsilon)\|_\infty/\varepsilon$ fluctuates around 0.3 for all choices of ε; for $d = 7$, this ratio begins at ≈ 0.1 for $\varepsilon = 0.1$, then decreases to ≈ 0.01 for $\varepsilon = 0.001$. This shows that our adaptive approximation algorithm works reasonably well. It appears slightly more effective in lower dimensions than in higher dimensions. A likely reason is that the underlying POSD structure can be more easily learned from a small pilot sample in lower dimensions than in higher dimensions.

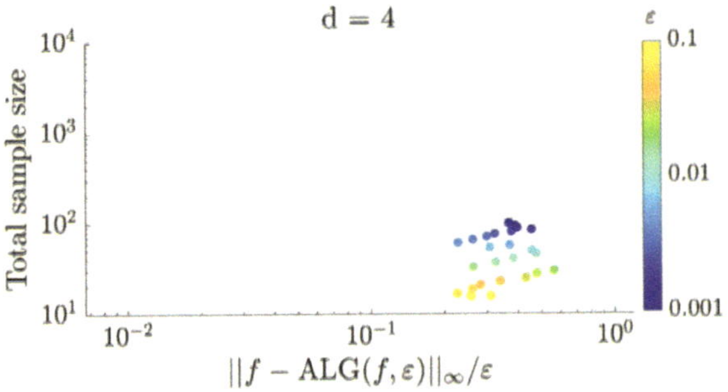

(a) f is a $d = 4$-dim. function with random Fourier coefficients.

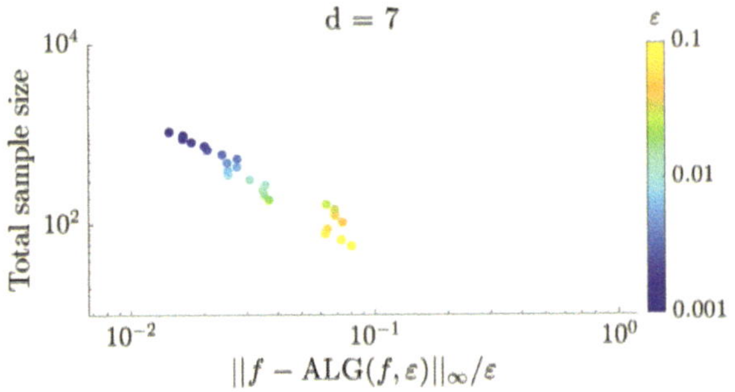

(b) f is a $d = 7$-dim. function with random Fourier coefficients.

Figure 7.1: Total required sample size as a function of error ratio $\|f - \mathrm{ALG}(f, \varepsilon)\|_\infty/\varepsilon$, with points colored by the absolute error tolerance level ε.

Bibliography

[1] J. Dick, F. Y. Kuo, Q. T. Le Gia, D. Nuyens and C. Schwab. Higher order QMC Petrov–Galerkin discretization for affine parametric operator equations with random field inputs. SIAM J. Numer. Anal. 52(6):2676–2702, 2014.

[2] F. J. Hickernell and L. l. A. Jiménez Rugama. Reliable adaptive cubature using digital sequences. In: R. Cools and D. Nuyens (eds), Monte Carlo and Quasi-Monte Carlo Methods: MCQMC, Leuven, Belgium, April 2014, Springer-Verlag, Berlin, pp 367–383, 2016, arXiv:1410.8615 [math.NA].

[3] F. J. Hickernell, L. l. A. Jiménez Rugama, and D. Li. Adaptive quasi-Monte Carlo methods for cubature. In: J. Dick, F. Y. Kuo, and H. Woźniakowski (eds) Contemporary Computational Mathematics — a celebration of the 80th birthday of Ian Sloan, Springer-Verlag, pp 597–619, 2018. 10.1007/978-3-319-72456-0.

[4] L. l. A. Jiménez Rugama and F. J. Hickernell. Adaptive multidimensional integration based on rank-1 lattices. In: R. Cools and D. Nuyens (eds), Monte Carlo and Quasi-Monte Carlo Methods: MCQMC, Leuven, Belgium, April 2014, Springer-Verlag, Berlin, pp 407–422, 2016, arXiv:1411.1966.

[5] P. Kritzer and H. Woźniakowski. Simple characterizations of exponential tractability for linear multivariate problems. J. Complex. 51:110–128, 2019.

[6] F. Y. Kuo, C. Schwab, and I. H. Sloan. Quasi-Monte Carlo finite element methods for a class of elliptic partial differential equations with random coefficients. SIAM J. Numer. Anal. 50:3351–3374, 2012.
[7] E. Novak and H. Woźniakowski. Tractability of Multivariate Problems Volume I: Linear Information. No. 6 in EMS Tracts in Mathematics, European Mathematical Society, Zürich, 2008.
[8] E. Novak and H. Woźniakowski. Tractability of Multivariate Problems Volume II: Standard Information for Functionals. No. 12 in EMS Tracts in Mathematics, European Mathematical Society, Zürich, 2010.
[9] E. Novak and H. Woźniakowski. Tractability of Multivariate Problems Volume III: Standard Information for Operators. No. 18 in EMS Tracts in Mathematics, European Mathematical Society, Zürich, 2012.
[10] A. Pietsch. Operator Ideals. Deutscher Verlag der Wissenschaften, Berlin, 1978.
[11] A. Pinkus. n-Widths in Approximation Theory. Springer-Verlag, Berlin Heidelberg, 1985, 10.1007/978-3-642-69894-1.
[12] J. F. Traub and A. G. Werschulz. Complexity and Information. Cambridge University Press, Cambridge, 1998.
[13] J. F. Traub, G. W. Wasilkowski, and H. Woźniakowski. Information-Based Complexity. Academic Press, Boston, 1988.
[14] A. G. Werschulz and H.Woźniakowski. A new characterization of (s,t)-weak tractability. J. Complex. 38:68–79, 2017.
[15] C. F. J. Wu and M. S. Hamada. Experiments: Planning, Analysis, and Optimization. John Wiley & Sons, 2009.

Radon Series on Computational and Applied Mathematics

Volume 26
Discrepancy Theory
Dmitriy Bilyk, Josef Dick, Friedrich Pillichshammer (Eds.), 2020
ISBN 978-3-11-065115-7, e-ISBN 978-3-11-065258-1

Volume 25
Space-Time Methods. Applications to Partial Differential Equations
Ulrich Langer, Olaf Steinbach (Eds.), 2019
ISBN: 978-3-11-054787-0, e-ISBN: 978-3-11-054848-8

Volume 24
Maxwell's Equations. Analysis and Numerics
Ulrich Langer, Dirk Pauly, Sergey I. Repin (Eds.), 2019
ISBN: 978-3-11-054264-6, e-ISBN: 978-3-11-054361-2

Volume 23
Combinatorics and Finite Fields. Difference Sets, Polynomials, Pseudorandomness and Applications
Kai-Uwe Schmidt, Arne Winterhof (Eds.), 2019
ISBN: 978-3-11-064179-0, e-ISBN: 978-3-11-064209-4

Volume 22
The Radon Transform. The First 100 Years and Beyond
Ronny Ramlau, Otmar Scherzer (Eds.), 2019
ISBN: 978-3-11-055941-5, e-ISBN: 978-3-11-056085-5

Volume 21
Hamilton-Jacobi-Bellman Equations. Numerical Methods and Applications in Optimal Control
Dante Kalise, Karl Kunisch, Zhiping Rao (Eds.), 2018
ISBN: 978-3-11-054263-9, e-ISBN: 978-3-11-054359-9

Volume 20
Fluid-Structure Interaction. Modeling, Adaptive Discretisations and Solvers
Stefan Frei, Bärbel Holm, Thomas Richter, Thomas Wick, Huidong Yang (Eds.), 2017
ISBN: 978-3-11-049527-0, e-ISBN: 978-3-11-049425-9

www.degruyter.com

Bei Fragen zur Produktsicherheit wenden Sie sich bitte an:
If you have any questions regarding product safety,
please contact:

Walter de Gruyter GmbH
Genthiner Straße 13
10785 Berlin
productsafety@degruyterbrill.com